LOVE AND INJUSTICE IN MEDICINE

LOVE AND INJUSTICE IN MEDICINE

Annotated Narrative Ethics Explorations

JEFF NISKER MD PhD

IGUANA

Copyright © 2022 Jeff Nisker
Published by Iguana Books
720 Bathurst Street, Suite 410
Toronto, ON M5S 2R4

All rights reserved. No part of this publication may be reproduced, stored in a retrieval system or transmitted, in any form or by any means, electronic, mechanical, recording or otherwise (except brief passages for purposes of review) without the prior permission of the author.

Publisher: Meghan Behse
Editor: Mariko Obokata
Front cover design: Jonathan Relph

ISBN 978-1-77180-609-1 (hardcover)
ISBN 978-1-77180-590-2 (paperback)
ISBN 978-1-77180-589-6 (epub)

This is an original print edition of *Love and Injustice in Medicine: Annotated Narrative Ethics Explorations.*

*It is only with the heart that one can see rightly;
what is essential is invisible to the eye.*

—Antoine de Saint-Exupéry, *The Little Prince*

To all my students from whom I have learned so much

Table of Contents

Introduction xi

Chapter Synopses xvii

Chapter 1: The Rotor 1

Chapter 2: "You Must Go to Medical School or Hitler Will Have Won" 9

Chapter 3: I'm Sorry Ronnie 17

Chapter 4: Philip 29

Chapter 5: Princess Margaret 45

Chapter 6: Miriam 51

Chapter 7: I'm Sorry Vaccine Came Too Late for You Janet 57

Chapter 8: Thank You Grace 65

Chapter 9: Beneath the Pagoda's Perch 72

Chapter 10: She Lived with the Knowledge 77

Chapter 11: Dr. King, The Little Prince, and Seeing with One's Heart in Medicine 87

Chapter 12: For Medical Students Protesting the Injustice of Clayoquot Sound 95

Chapter 13: The "Helix of Life" Revisited: DNA in Concrete and Not 98

Chapter 14: Beneath the BMW's Wheels 121

Chapter 15: The Injustice of Needing Angelina Jolie 130

Chapter 16: A Brief and Personal History of "What's in a Name" 142

Chapter 17: Ruth 164

Chapter 18: Victor 175

Chapter 19: Canadian COVID Injustice On Beaches and Beach

              Volleyball Courts.. 183

Chapter 20: Webinar Physicians' Cavalier Terms for

              COVID-Ventilator Triage of Disabled Persons.................... 204

Chapter 21: COVID Injustice Before I Heard the Word "COVID" 207

Chapter 22: COVID Aggression Condemns a Muslim Family

              Near Our Medical School.. 214

Chapter 23: The Lottery 217

Chapter 24: Antivaxxer Xenophobic COVID Violence 221

Chapter 25: The Arrogance of "But All You Need Is a Good

              Index Finger" 228

Chapter 26: Our Third COVID Summer 245

Appendix I: The Psych Experiment 251

Acknowledgements.. 259

About the Author... 261

Introduction

Love and Injustice in Medicine: Annotated Narrative Ethics Explorations interrogates the injustices I have witnessed amidst the gift of being a physician. The linked short stories and prose poems, though deeply personal, are generalizable to the injustices increasingly prevalent in countries where single-tiered health systems are evaporating. *Love and Injustice in Medicine* expresses my regret for not having better shared the gift of my years in medicine by better taking on the systems-problems that increasingly ignore socio-economically disadvantaged persons; systems-problems that were uncovered and augmented by the COVID pandemic. *Love and Injustice in Medicine* enlists the power of creative non-fiction narratives to juxtapose readers with the persons immersed in the vortex of both health problems and health-systems problems, too-often without a social-justice life jacket.

An explanation for my coming to the position that creative non-fiction narratives are the best way to immerse health professionals, students, clinicians, and the general public in the position of persons who require health promotion and care is offered in Chapter 11, "Dr. King, The Little Prince, and Seeing with One's Heart in Medicine." In this chapter I recount reading to my pajamaed children beneath a poster of Dr. Martin Luther King Jr. the words of The Little Prince's friend the wise Fox, "It is only with the heart that one can see rightly; what is essential is invisible to the eye."[1]

The epiphany of the wise Fox's words combined with the compassion in Dr. King's eyes made me a better physician, a better educator, a better father, as well as a better person, and a better contributor to compassionate health-policy development. I have observed the power of the Fox's words when using creative non-fiction in small classrooms, and in large auditoria. I have also witnessed the power of the Fox's words in public theatres following productions of my social-justice plays.[2] These plays were described in a 2020 *Canadian Medical Association Journal* "Profile" by Miriam Shuchman as "Theatre of Social Justice."[3] Writing creative non-fiction narratives for the purpose of imbuing social justice in the hearts of current and future health professionals, and other health-policy makers, not to mention the general public who through their votes are also health-policy makers, has been my narcotic over years of multiple disc-herniations, and has more recently been my comfort food during chemotherapy.

Creative non-fiction narratives can help humanity emerge as the primary imperative in health-policy development through fostering empathy for persons too often invisible at the health-policy table, and when visible too often appearing too different from the other persons at the table. Creative non-fiction narratives can encourage us to see the uniqueness and beauty of each person,[4] and thus promote compassion for the individual person, and by extension for their community. Stories that promote compassion can promote the social justice that insists on improving the social determinants of health[5] for

---

[1] de Saint-Exupéry, 1943.
[2] The plays include *Sarah's Daughters* (Nisker, 2012), *Orchids* (Nisker, 2001b, 2012), *Calcedonies* (Nisker, 2010, 2012); A Child on Her Mind (Nisker & Bergum, 2007; Nisker, 2012); Camouflage (Nisker, 2012); Philip 2003; Nisker, 2012).
[3] Shuchman, 2020.
[4] Nisker, 2001b, 2012, 2015b.
[5] Frazee, Gilmour, Mykitiuk, & Bach, 2002; Mykitiuk & Nisker, 2010; Siegler & Epstein, 2003; World Health Organization, 2010.

all members of the community, not just for persons who are financially better off and thus already have better access to health through better nutrition, better physical environments, better social environments, and better access to health professionals, as well as better access to other social determinants.[6]

Love and Injustice in Medicine begins with my head confined inside an MRI machine's rotating magnet during the acute phase of a stroke. *Love and Injustice in Medicine* then reminisces back to my experiences with medicine, from being told "You Must Go to Medical School or Hitler Will Have Won,"[7] through my years as a medical student, resident, and cancer-research fellow, to my years as a clinician-scientist, health-ethics researcher, and Chair of national and international health-policy committees. It was particularly during my years on health-policy committees that I could have made a greater contribution to equality in health promotion, and thus could have better contributed to diminishing the disparity in suffering later observed during the COVID-19 pandemic.

Several of the chapters in *Love and Injustice in Medicine* address the absence of imperative for physician-advocacy,[8] not only for persons in our practices, but for persons in our communities, in our countries, and globally. I contend the physician's "role of advocate"[9] must supersede "gatekeeper/manager" on the list of physician "roles."[10] Indeed, "gatekeeper/manager" should not be a role at all for physicians in their practices, but instead a role for hospital and government committees, on which physicians, in my opinion, have the obligation to serve. These committees must also include and be responsible to members of diverse communities, in order to ensure

---

[6] Frazee, Gilmour, Mykitiuk, & Bach, 2002; Mykitiuk & Nisker, 2010; Siegler & Epstein, 2003; World Health Organization, 2010.
[7] See Chapter 2, "You Must Go to Medical School or Hitler Will Have Won."
[8] See Chapter 3, I'm Sorry Ronnie; See Chapter 14, Beneath the BMW's Wheels.
[9] Royal College of Physicians and Surgeons of Canada, 2021.
[10] Royal College of Physicians and Surgeons of Canada, 2021.

equal access to best practices in health promotion and care is preeminent in health-policy making.

Many of the injustices I have witnessed have been amplified by the increases in private option in healthcare, overlooked purposefully by supposedly social justice–based governments. As Canada has one of the lowest physician-patient ratios of "developed" countries,[11] an increase in physician time-devotion to persons who can pay a premium for physician attention will increase the percentage of persons lacking access to a family physician,[12] and dramatically lengthen wait times to be seen by specialists.[13]

Preferential access for the financially privileged already exists in Canada, albeit illegally, in a black-market system in which persons can purchase guaranteed access to a family physician immediately, access to a specialist within a week,[14] as well as imminent access to diagnostic and surgical facilities. This injustice of preferential access concomitantly tramples the access of persons who are socio-economically disadvantaged,[15] disproportionately including persons with disabilities, persons of Indigenous heritage, and recent immigrants.[16] This injustice will increase dramatically should the current threat of dissolution of Canada's single-tiered health system become more than a threat.[17]

---

[11] Nisker, 2018, 2019; World Bank Group, 2018.
[12] Statistics Canada, 2017.
[13] Nisker, 2019.
[14] Cribb & Isai, 2017; Silversides, 2008.
[15] Nisker, 2019.
[16] Burlock, 2017; Gaetz, Donaldson, Richter, & Gulliver, 2013; Mikkonen & Raphael, 2010; Nisker, 2019; Preibisch & Hennebry, 2011; Smylie, Firestone, Spiller, & Tungasuvvingat Inuit, 2018.
[17] This threat has worked its way through the lower and superior courts in the province of British Columbia (Cambie Surgeries Corp. v. British Columbia (Medical Services Commission), 2010; Nisker, 2019), and will soon be heard by the Supreme Court of Canada. If the Supreme Court of Canada allows this threat to become a reality, more physicians will choose to "opt out" of the public system for the greater financial rewards of a private system.

Love and Injustice in Medicine attempts to foster discomfort under the spectre of the dissolution of Canada's single-tiered health system, as I worry that without a social-justice imperative, the humanity of persons in health-policy deliberations will dissolve in the red ink of "accountability" to the black ink of "quantifiable parameters." These too heavily applauded quantifiable turnstiles are too heavy to rotate for the less quantifiable parameters of equality and compassion;[18] parameters lacking emphasis in balanced-budget imperatives.

Several of the chapters in *Love and Injustice in Medicine* explore the injustices I am witnessing in genetics-related medicine,[19] including lack of funding for genetic counselling, testing, and surveillance strategies for persons at high risk of hereditary cancer.[20] I recount the impact of these funding-lacks on my family,[21] on my partner's family,[22] on close friends,[23] and on myself.[24] The benefits of genetic screening for prevention and early detection of cancer stand in contrast to the discrimination prenatal genetic screening may foster against persons living with genetics-related disabilities; persons who may be cruelly seen in this time of burgeoning genetics technologies as persons whose disability could have been prevented by preventing their birth.[25]

---

[18] Nisker, 2012, 2015b.

[19] See Chapter 10, She Lived with the Knowledge; See Chapter 13, The "Helix of Life" Revisited: DNA in Concrete and Not; See Chapter 18, Victor; See Chapter 25, The Arrogance of "But All You Need Is a Good Index Finger."

[20] Joseph, Rab, Panabaker, & Nisker, 2015; Nisker, 2007, 2012; See Chapter 10, She Lived with the Knowledge; See Chapter 15, The Injustice of Needing Angelina Jolie.

[21] See Chapter 3, I'm Sorry Ronnie; See Chapter 10, She Lived with the Knowledge.

[22] See Chapter 18, Victor.

[23] See Chapter 25, The Arrogance of "But All You Need Is a Good Index Finger."

[24] See Chapter 25, The Arrogance of "But All You Need Is a Good Index Finger."

[25] Mykitiuk & Nisker, 2010; Nisker, 2001b, 2012; Nisker, Baylis, Karpin, McLeod, & Mykitiuk, 2010; Vanstone, King, de Vrijer, & Nisker, 2014; Vanstone, Yacoub, Winsor, Giacomini, & Nisker, 2015.

Several of the chapters in *Love and Injustice in Medicine* address the injustices I am witnessing during the COVID pandemic as our health system is stretched beyond its budgeted capacity; a stretch easy to accommodate with a slight increase in taxes on corporations and wealthy citizens. COVID has become perhaps the most important medical and social-justice challenge in history, magnifying the lack of access and equality in health and social systems. COVID has become the elephant in the room, albeit on Zoom, no matter what health policy we deliberate over, no matter what seminar we facilitate.[26]

COVID has been a time I have been sentenced to the sidelines[27] because of my chemo-related impaired immunity, and my colleagues' concern that COVID will claim me as a casualty. I continue to try to be permitted to enter the fray as 2022 progresses and *Love and Injustice in Medicine* goes to press. It is urgent and imperative that creative non-fiction narratives convey the position of persons threatened by diminishing access to health promotion and care; persons who will be in greater jeopardy should our single-tiered health system dissolve in neoliberal-thinking ink.[28] I plan to use *Love and Injustice in Medicine* in my participation in a factum for the Supreme Court of Canada when the threat to our single-tiered health system reaches its hallowed shore.

I hope you will find these creative non-fiction narratives useful for yourselves and your communities, and that they will assist you to help ensure future health policies will insist on social justice.

---

[26] See Chapter 20, Webinar Physicians' Cavalier Terms for Triage from COVID Ventilators of Persons with Disabilities.
[27] Nisker, 2022
[28] Nisker, 2019.

Chapter Synopses

The Rotor is a short story that evolves from my spinning thoughts inside an MRI machine's rotating magnet during a cerebellar stroke. I had declined a blindfold like a courageous resistance fighter facing a firing squad, as I wanted to look brave in front of our Hospital's radiology technicians. However, I forgot to close my eyes, which was a mistake considering my claustrophobia,[29] and had to non-courageously press the "panic button" for the technicians to quickly conveyor-belt me out of the tight tunnel. After I was rolled out of the magnet, one of the technicians soothed my hand while shaking her head at a physician who should have known better. When my claustrophobia had calmed a bit, she looked into my eyes, and said in a trying-to-calm-me voice, "You at least have to close your eyes," mercifully leaving out, "You fool." She smiled before the conveyor belt ominously rolled back into the magnet's massive machinery.

My 45 minutes of immobility in the magnet reminded me that I had not yet written the story I had promised the woman I call Ruth,[30] who after driving her chin-operated powerchair into my knees demanded, "Hey Doc who writes stories, write one about me and the fucking way your system treats me." Ruth died from inadequate

---

[29] I revisit my claustrophobia in Chapter 17, Ruth.
[30] See Chapter 17, Ruth.

health and social support soon after she began describing to me the injustices inflicted upon her by Canada's supposedly wonderful health and social support systems.[31] It took me several years to "find the time" to fulfill my promise to Ruth, and I only found the time when I was banished from our Hospital because my immune system had been shut down by chemotherapy.

Medicine begins for me with my Uncle's quiet declaration at our kitchenette table, "You Must Go to Medical School or Hitler Will Have Won." Although it has been a gift being a physician, I had been groomed from birth by my maternal Grandmother[32] to be a human rights lawyer, to pursue the social justice her imminent death from early onset breast cancer would not permit her to pursue. No plan for me to go to medical school had been mentioned previously, and I was firm in my aspiration to become Atticus Finch.[33]

This narrative reminisces the motivations for my different imperatives for my future of my Grandmother and Uncle. My Grandmother had been an indentured servant in the Ukraine, and a social-justice advocate in Toronto, until early onset hereditary breast cancer claimed her.[34] She wanted me to do good in the world by speaking for socio-economically disadvantaged persons who generally have no voice.[35] My Uncle had escaped Poland, along with

---

[31] My tributes to Ruth include my play *Calcedonies* (Nisker, 2012), and my novel *Patiently Waiting For...* (Nisker, 2015b).

[32] I refer to my Grandmother, my second mother and social-justice mentor, in several chapters. See Chapter 5, Princess Margaret; See Chapter 13, The "Helix of Life" Revisited: DNA in Concrete and Not.

[33] Atticus Finch is the lawyer in Harper Lee's novel *To Kill a Mockingbird*. Gregory Peck won an Academy Award for his portrayal of Atticus Finch in the film version directed by Robert Mulligan.

[34] See Chapter 5, Princess Margaret; See Chapter 13, The "Helix of Life" Revisited: DNA in Concrete and Not.

[35] Speaking for those who have no voice is a biblical reference in the Book of Solomon related to King Lemmens.

my paternal Grandmother, my Father and my Aunt, just prior to Hitler's invasion.[36] Grandfather, the town's only doctor, had made a pact with the town's priest and its mayor that if they forged papers making his daughter and her children Christians, so they would be permitted to emigrate to America, he would remain to provide for the town's medical needs. He remained until he and the rest of our family were transported to Auschwitz. My Uncle's reason for me to go to medical school was it was a family "tradition" for someone in each generation to be a physician, and my Uncle felt that I was the only one in my generation who could get into medical school.

Although it was my Uncle who said, "You must go to medical school or Hitler will have won," the words that ultimately gave me the gift of being a physician were my Father's, "It will be easier for you to change the world with MD after your name," along with my Mother's emphasis, "You will still be able to do what *Bubie*[37] wanted you to do if you're a doctor, and you'll be able to help people in other ways; maybe you'll even find a cure for breast cancer."[38]

I'm Sorry Ronnie is my *mea culpa* to the youngest in our small generation of cousins. We were a tight family, as immigrant families tend to be, especially families with Holocaust histories. Ronnie was "different" from the rest of us in that instead of using the language of words, Ronnie used the language of smiles, hugs, and other physical expressions. Ronnie entered puberty around the time I entered medical school. Ronnie developed a strong and agile athletic body that neighbours began fearing, especially if they had daughters. They encouraged my Aunt to send Ronnie away, an all-too-common

---

[36] See Chapter 3, I'm Sorry Ronnie.
[37] *Bubie* is the Yiddish word for grandmother.
[38] My Grandmother had died recently from early onset breast cancer, and my Mother would also die of breast cancer 10 years after she said this to me. I still have not reconciled the injustice their too-young deaths. Also See Chapter 10, She Lived with the Knowledge.

sentence for persons of difference at that time. Ronnie's doctor eventually concurred, and Ronnie was sent to the infamous Huronia,[39] originally named The Orillia Asylum for Idiots.

During medical school, I regretted not visiting Ronnie more frequently at Huronia, a dark and sinister place. Ronnie had no words to describe further what he felt or experienced. When I did visit Ronnie, he would always run to me, smile at me, embrace me in his warm arms, and eagerly show me the long rows of institutional beds in his room, the long rows of institutional tables in the dining hall, the caged windows, the white-uniformed staff. When I moved two and a half hours farther away to do my residency, visits to Ronnie were diminished. When I became a parent, Ronnie's predicament seemed to fade from my mind, until my Father called to tell me Ronnie was dead. Ronnie had fallen off the back of the institution's sleigh during a group ride over the supposedly frozen Lake Couchiching. The next day a search team found Ronnie's footprints heading back to Huronia, until they abruptly ended in a hole in the ice. This pallbearer at Ronnie's funeral observed water dripping out of Ronnie's coffin, as if Ronnie was still speaking to me; speaking as always without words.

In **Philip** I retrace the footsteps of a precocious 12-year-old boy from when he follows his parents through "The Atrium" of Toronto's Hospital for Sick Children, through the floor on which Philip will receive new prescriptions, through the decisions being made for him by his physicians, his parents, and the tradition of medicine. "Philip" surfaces the importance and complexity of informed choice in decision-making through the eyes of a child developing at an intellectually accelerated rate. It surfaces the imperative to afford the dignity of self-determination whether the person is young or old, or whether the person requires assistance in decision-making or has full

---

[39] Ballingall, 2016.

capacity to make decisions, particularly if the decisions are not consistent with the wishes of medicine.

"Philip" remains a beacon for me regarding the requirement of the principle of justice in informed choice in decision-making. I feel I owe it to Philip to give him a voice in medicine; a voice in our learning, a voice in our decision-making, a voice that Philip did not have until his story nears its end.

As a senior medical student, I "rotated" to Princess Margaret Cancer Hospital[40] for my last two weeks on Paediatrics service. Although I found all eight weeks on Paediatrics tough because of the injustice of children being so ill, I was not prepared for the injustices I witnessed at "Princess Margaret." As cancer is mainly an adult disease, "Princess Margaret" was mainly an adult hospital, void of the colourful helium balloons and the animated wall-decals of the children's hospital where Philip had previously undergone treatment.[41] When I was a medical student, many children with cancer had to have some of their treatment at Princess Margaret because that was the location of the massive radiation machines.

The bare heads of the six- to ten-year-old boys in my "Ward" bore blue plus signs, tattooed as a target for radiation to their brains. Their eyes bore dark circles. Although I had been instructed in medical school to "limit emotion," to always be "objective," to never let my heart get in the way of good medicine, I knew in my first minute at Princess Margaret that my heart was going to get in the way, and that I was going to let it. To this day, when I reflect on the young persons caught in the injustice of cancer's wrath, I am not able to accept the continued impotence of my profession in achieving 100-percent cures.

---

[40] In 1998, Princess Margaret Hospital became part of the University Health Network, and in 2012 it was renamed Princess Margaret Cancer Centre.
[41] Nisker, 2012.

I was "Chief Resident" when I met Miriam, a 39-year-old woman referred to my mentor by a physician in Toronto. The purpose of the referral was to discourage Miriam's insistence that her ovaries needed to be removed to prevent non-hereditary cancer. My mentor could not talk her out of the surgery, so he asked me try. Miriam forced a smile before saying, "I know you're going to think I'm crazy, but I know I'm going to die soon if you don't take out my ovaries." I gave Miriam a reassuring smile, pulled curtains around her bed, and asked her to please tell me her family's cancer history. Instead Miriam handed me a wrinkled sheet of yellowing paper illustrating a sinister family tree. Ovarian cancer was not supposed to be hereditary, but Miriam's family's ovarian cancer was. Miriam explained that her family tree's trunk was short because the generation above hers had perished at Auschwitz.[42] I reassured Miriam that we would not let her die from ovarian cancer.

About an hour later, a medical student ran up to me while I was making evening rounds. He was trembling as he told me he "found something" in Miriam's left breast when doing her pre-op physical. As I re-entered Miriam's room, she was breathing rapidly, pleading, "Please God no, please God no, please *Gotenu* no." On examining Miriam's left breast, and the lymph nodes under her left shoulder, it was clear that Miriam would not die from ovarian cancer. It would be another decade before BRCA gene mutations[43] were found to be a cause of both early onset breast cancer and ovarian cancer.

I'm Sorry Vaccine Came Too Late for You Janet reminisces a 29-year-old woman referred to my mentor from the North. Janet's body

---

[42] See Chapter 2, "You Must Go to Medical School or Hitler Will Have Won."
[43] BRCA gene mutations were discovered in 1990s, in part by the Canadian physician and cancer researcher Dr. Steven Narod (Narod, Brunet, Ghadirian, Robson, Heimdal, *et al.*, 2000; Narod, Feunteun, Lynch, Watson, Conway, *et al.*, 1991; Narod, Lynch, Conway, Watson, Feunteun, *et al.*, 1993) at the University of Toronto; Nisker 2007; Nisker 2012; Nisker 2013.

was emaciated from a large cervical[44] cancer's ravenous appetite; emaciated to the point that she looked like she was in Auschwitz.[45] Ironically, I had to deny Janet food, even oral fluids except for ice chips, in order to "prep" her bowel for surgery, in anticipation of having to remove a significant section of her bowel if attached to the cervical cancer. I administered nutrition by giving Janet a new protein and vitamin mixture through a large bore intravenous, which I had previously inserted under her right collarbone. Twice a day, as I checked to make sure the "hyperalimentation" insertion site was not becoming infected, I engaged Janet in conversation, and grew to know her better than any other patient during my residency; perhaps better than any patient to this day.

Three weeks after I met Janet, I rolled her bed into the OR, where my mentor and I would work for hours trying to remove Janet's tumour "in one block." This strategy was essential to prevent the spillage of cancer cells that would doom Janet. After we removed Janet's cancerous uterus, hugely enlarged by the cancerous cervix, we stared down at a centimetre-size flake of cancer still attached to the side wall of Janet's pelvis. My mentor picked at the cancer with fine-point tweezers, and scraped at it with his scalpel before walking away from the operating table. He then pounded his gloved fists against a wall, before resting his head against it. He ripped off his gloves, threw them to the floor, said "close Jeffer," and left. It would be up to me when Janet's anaesthesia dissipated to explain what we could not achieve.

Thank You Grace reminisces the gift of a woman of faith who was transferred to my care from "upcountry", hemorrhaging after her

---

[44] The cervix is the lower part of the uterus protruding into the vagina. I have written previously at length about BRCA gene mutations, and their preponderance for causing breast and ovarian cancer in women of Jewish heritage, including She Lived with the Knowledge, Chapter 10.

[45] See Chapter 2, "You Must Go to Medical School or Hitler Will Have Won."

sixth birth. It was my first weekend on call as a graduated specialist, and all my colleagues had vacated to attend our Annual National Conference. When the ambulance transporting Grace alerted me it was near, and that its patient was near death, I crashed down four flights of steps to await Grace. As she was stretchered through the ER doors, her light blue eyes embraced me, and her serene smile soothed me. I immediately asked a nurse to "type and cross" her for four units of blood, but Grace softly said that she could not accept blood. Grace's uterus was up to her ribs, likely because her uterine muscle had lost its tone and her "upcountry" physician had stuffed her uterus with cloth packing to try to stop Grace's hemorrhage. The packing was a wrong decision because of imminent infection, but the packing saved Grace's life.

I took Grace directly to the ICU, where its physician refused to accept Grace because Grace refused to accept blood. Eventually he relented, and Grace was transferred to an ICU bed, and its head dipped down to encourage more blood to flow to her brain. Grace's cognition continued to be translucent as we conversed through the night. By dawn it became clear that Grace was more worried about me than herself for several reasons, including that I did not accept Jesus. The packing needed removing or it and Grace would become septic. This chapter continues with Grace's operating room experience.

Beneath the Pagoda's Perch began on my way to speak at a conference at the Center for Literature and Medicine in Ohio, when I observed a highway sign indicating the turnoff to the "Town of Kent." This sign reminisced the medical-student me staring in disbelief at television images. As soon as I arrived at the conference, I started asking faculty if they would pilgrimage with me to Kent State University. One of the faculty agreed, quietly adding she had been a student at Kent State in 1970, and had personally witnessed the carnage of May 4.[46] When we arrived at Kent State, she solemnly led me to the hill overlooking the site of the massacre. The hill was still topped by the pagoda-like rain shelter violently engraved in my brain. I fell to my knees, barraged by afterimages of evening news footages still inconceivable to me, but all too real for her.

For the 25th anniversary of the Kent State massacre, I was asked to write a poem to be posted on the "In Memoriam" wall that had been erected on the Kent State campus. I wrote "Beneath the Pagoda's Perch."[47] As I was writing this commemorative poem, I appreciated that I was also commemorating the day I first felt the foreboding that our world would never be the sanctuary of justice that many of my generation had been confident we could achieve. My participation in the 50th anniversary of the Kent State massacre on May 4, 2020, dissolved in COVID precautions.

She Lived with the Knowledge began with a request from the organizers of a Cancer Care Ontario Annual Conference to write a poem to be read at the opening ceremonies. "She Lived with the Knowledge" poured from my heart onto tear-dripped paper. The poem was passionately read by a tear-dripped nurse educator at the Conference, and also read at many other conferences and educational events. "She Lived with the Knowledge" explores the injustice of early onset breast cancer through my love for my Mother and Grandmother, and my fear for my Sister.

At the time of my writing "She Lived with the Knowledge," most physicians in Canada did not know about BRCA gene mutations, though the mutation had been discovered by a Canadian physician[48] and his collaborators. Thus most women at risk of early onset breast

---

[46] Gordon, 1990.
[47] Nisker, 1995.
[48] Narod, Brunet, Ghadirian, Robson, Heimdal, *et al.*, 2000; Narod, Feunteun, Lynch, Watson, Conway, *et al.*, 1991; Narod, Lynch, Conway, Watson, Feunteun, *et al.*, 1993.

cancer in Canada lacked choice regarding accessing genetic counselling and testing, as well as the increased surveillance and prevention strategies that could mitigate them joining the several thousand young women who die each year from early onset breast cancer in Canada.[49] Clearly public outreach was required, so I expanded "She Lived with the Knowledge" into the full-length play *Sarah's Daughters*,[50] which eventually toured Canada and four other countries to promote awareness of BRCA gene mutation breast cancer risk. Yet it took an American woman, Angelina Jolie, going public with her story[51] to educate Canadian women and their physicians regarding BRCA gene mutations and the hereditary risk of breast and ovarian cancer.[52]

Dr. King, The Little Prince, and Seeing with One's Heart in Medicine recounts the epiphany that led me to use creative narratives to imbue compassion and social justice in medical students and physicians. Every night my pajamaed children huddled with me in the hall between their bedrooms beneath a framed poster of Dr. Martin Luther King Jr. and excerpts of his "dreams" spoken from Abraham Lincoln's marble feet at the conclusion of the March on Washington.[53] As Dr. King stared down on us, we sang the social-justice songs of Bob Dylan and Buffy Sainte-Marie, and read the social-justice writings of Gandhi and Lao Tzu, and of course portions of the passionate speeches of Dr. King. We called these intimate moments "Thoughts for the Night," a termed coined by my middle son's frequent-sleepover friend. This narrative strategy of imbuing compassion and social justice in my

---

[49] Joseph, Rab, Panabaker, & Nisker, 2015; Nisker, 2013.
[50] Nisker, 2012.
[51] See Chapter 15, The Injustice of Needing Angelina Jolie.
[52] Joseph, Rab, Panabaker, & Nisker, 2015.
[53] History.com, 2010.

children was different from the method used by my Grandmother,[54] but not by much.

When sharing the concept of "Thought for the Night" over dinner at a medical conference, a colleague asked if I had ever read my children *The Little Prince*. I had no knowledge of the existence of this book, and was surprised to find several copies in the airport bookstore the next day. *The Little Prince* was not like anything we had read previously. Though my children were enthralled by the story and the paintings, I was lukewarm about both until I read the line in which the wise Fox says, "It is only with the heart that one can see rightly; what is essential is invisible to the eye."[55] I must have reread this line several times because my eldest asked, "Are you okay, Dad?" I responded, "It's time to go to bed." I needed to reflect on the line that continues to echo through my mind to this day; the line I hope will protect medical hearts from succumbing to the numbing of medical education's objectifications that diminish our ability to be compassionate physicians.

For Medical Students Protesting the Injustice of Clayoquot Sound was written for a group of second-year medical students who had asked me to go with them to be chained to one of Clayoquot Sound's massive trees to protest their "clear-cutting." I had to decline their invitation because of clinical responsibilities, though like our medical students, I was abhorred by the television images of the clear-cutting of Clayoquot Sound's old-growth forests. As part of my apology for disappointing these social justice-inspired students, I promised to write a 60s-style protest song they could sing while they were chained to the trees, and while they were being arrested. I also promised I would get them out of jail.

---

[54] See Chapter 2, "You Must Go to Medical School or Hitler Will Have Won"; See Chapter 5, Princess Margaret; See Chapter 10, She Lived with the Knowledge; See Chapter 13, The "Helix of Life" Revisited: DNA in Concrete and Not.
[55] de Saint-Exupéry, 1943.

The medical students and I saw injustice in clear-cutting of trees, and in how this injustice promoted injustice in our treatment of humans. The First Nations peoples of Clayoquot Sound saw clear-cutting as beyond injustice; it was sacrilegious. The gift of Clayoquot Sound was further realized in 2018, when I paddled across Clayoquot Sound to Meares Island with a First Nations guide and experienced a spiritual place beyond my imagination. I thank my medical students for sending me there. Although the carnage of Clayoquot Sound had stopped, we witnessed on television in 2021 the slaughter of the old-growth forests just south of Clayoquot Sound.[56]

The "Helix of Life" Revisited: DNA in Concrete and Not arose from the multi-storey DNA molecule planted in front of the main entrance to the University of Toronto Medical School's brand-new building when I was a first-year student there. The cement molecule was planted to encourage the solid-science future of genetic research; however, the "Helix of Life" soon became the site of my first social-justice protest as a medical student. The protest was intended to prevent the official opening of our medical school's new building by our province's premier, who was being pummelled in the press for having rewarded a major election campaign donor with a tender-free construction contract for the Ontario Power Building[57] rising across the road from our Medical School.[58] Just before our premier was set to cut the red ribbon on the semi-transparent sheath shielding the "Helix of Life" from snow, one of his aides noticed the two giant snowballs we had placed in the "anatomically correct" position on either side of the sheathed "Helix."

---

[56] Azpiri, 2021.
[57] Paikin, 2016, p. 172.
[58] Over the next few years, that Premier proved to be quite progressive, especially for a Conservative Premier, particularly in regard to Ontario's education system. He even started an educational television channel called TVOntario (Paiken, 2016).

Forty years later, I revisited the "Helix" just as Canada's *Genetic Non-Discrimination Act*[59] was before Parliament. I had been asked to speak at the University of Toronto, and as usual went on my ritual jog through campus and beyond, circling back to the "Helix of Life." However, I now found the DNA spiral represented to me the spiralling-out-of-control applications of DNA screening of embryos and fetuses. Although Canada's *Genetic Non-Discrimination Act*[60] legislates against discrimination in employment and insurance, the *Act* does not address the genetic discrimination that DNA screening of embryos and fetuses insidiously fosters against persons living with genetics-based disabilities that could have been prevented by preventing their birth.[61] This chapter insists we must work for further anti-genetic discrimination legislation, and argues that we are all less when our worth is determined by genetics.

Although the man **Beneath the BMW's Wheels**[62] had been haunting me for many years, I was stimulated to finish this prose poem by COVID's acceleration of the plight of socio-economically disadvantaged persons in Canada. "Beneath the BMW's Wheels" reflects my inadequacy of advocacy for the health of a homeless man who reached out to me from where he had sheltered under a BMW in the alleyway I was cutting through on my way to a research meeting at the University of Toronto. Stopping to assist this man, I was blown away by the poverty in the alleyway juxtaposed to the luxury vehicles parked in this newly gentrified neighbourhood.

---

[59] Government of Canada, 2017.
[60] Government of Canada, 2017.
[61] Nisker, 2001b, 2012; Nisker & Gore-Langton, 1995; Vanstone, Cernat, Nisker, & Schwartz, 2018; Vanstone, King, de Vrijer, & Nisker, 2014; Vanstone, Yacoub, Winsor, Giacomini, & Nisker, 2015.
[62] *Beneath the Wheel* is a Hermann Hesse novel exploring his suppression under an archaic Swiss education system in the early 1900s. I borrow from Hesse's title to recount the injustice inflicted on the socio-economically suppressed persons in Canada amidst copious wealth.

"Beneath the BMW's Wheels" explores the obligation of health professionals to advocate for persons beyond those in our practices, including persons on our streets and in remote regions of our country. "Beneath the BMW's Wheels" reaches out to the once-idealistic medical students, who have often become less-than-idealistic physicians, in the hope we will come together to promote the health of socio-economically disadvantaged persons during and after the COVID pandemic. No person should ever be left behind beneath a BMW's wheels.

The Injustice of Needing Angelina Jolie echoes back to Canada's shame of requiring an American woman to go public with her story to raise awareness of BRCA gene mutation–related breast and ovarian cancer. Angelina Jolie provided the outreach to young Canadian women at high risk that Health Canada and Canadian health-policy pundits had not vigorously pursued. She became aware of her familial risk after her mother's death at age 56 brought to light a BRCA gene mutation inheritance pattern.[63]

Within a week of Angelina Jolie "going public," we observed a tripling in referrals to our Cancer Genetics Clinic, including many from physicians who had never previously referred.[64] Without Angelina's story, many of these women would not have had the opportunity to choose genetic counselling, testing, increased surveillance, and prevention strategies. Angelina's story brought to light the inexcusable 10-year delay in Canada's initiation of the opportunity of BRCA gene mutation–testing for women at high risk when compared with European countries, and even the United States for persons through Medicare, HMOs, and private

---

[63] BRCA gene mutations are autosomal dominant and of high penetrance. Autosomal dominant means you only need to inherit the gene from one parent to potentially develop the related condition; high penetrance means that if you do inherit the gene, you are very likely to develop the condition.
[64] Nisker 2013; Joseph et al., 2015.

insurance. Angelina's story has now prevented the deaths of approximately 10,000 young Canadian women whose physicians had been ignorant of BRCA gene mutations.[65] Angelina Jolie concluded her story with, "Life comes with many challenges. The ones that should not scare us are the ones we can take on and take control of."[66]

A Brief and Personal History of "What's in a Name"[67] interrogates terms used in reproductive genetics, building on Juliet Capulet's claim, "That which we call a rose by any other name would smell as sweet."[68] Scientists and clinicians, including myself, have used sweet-smelling names, unknowingly and also knowingly, to encourage the acceptance of complex cutting-edge genetic research with non-viable human embryos. Sweet-smelling names continued in clinical research to spray perfume for research ethics boards, research-funding agencies, government regulators, hospital administrators, and eventually the general public. The sweet-smelling names include "pre-embryo," "suitable embryo," "healthy embryo," "non-invasive prenatal testing," and "donation." Most recently, "mitochondrial replacement therapy" is the perfumed term used to camouflage germ-line nuclear transfer, a prohibited activity in anti–reproductive cloning legislation in most countries,[69] including Canada's *Assisted Human Reproduction Act*.[70]

It is essential that transparent rather than camouflaged terms are used to ensure informed choice occur for persons who come to clinicians for information regarding reproductive genetics. It is similarly essential that transparency occur for research ethics boards,

---

[65] Nisker, 2013; Joseph, Rab, Panabaker, Nisker 2015.
[66] Jolie, 2013.
[67] Nisker, 2020b; Shakespeare, 1734.
[68] Nisker, 2021a; Shakespeare, 1734.
[69] Nisker, 2015a, 2021a.
[70] Government of Canada, 2004.

research-funding agencies, government regulators, hospital administrators, and the general public. Indeed, transparent terms are essential for informed decision-making regarding all existing and future medical encounters.

Ruth is a work of fiction built on true occurrences in the life of the woman who drove her chin-operated powerchair into my knees outside my office, demanding, "Hey Doc who writes stories, write one about me and the fucking way your system treats me." Although I had no choice but to make this promise immediately, the time commitment of my clinical and research work, not to mention being a parent, distanced Ruth to the back of my mind until I found myself immobile in an MRI machine's tight tunnel during the acute phase of a stroke.[71] In the claustrophobic[72] tunnel, I gathered my thoughts for the story I promised Ruth by writing a prose poem, as is my practice for all my narratives. Over the next few months, I developed the poem into the short story "Calcedonies"[73] for the *Canadian Medical Association Journal*, and later expanded "Calcedonies" into a full-length theatre version,[74] and eventually the novel *Patiently Waiting For...*.[75] The original prose poem opens "Ruth" in *Love and Injustice in Medicine*.

"Ruth" begins a discussion of the injustices inflicted upon persons with disabilities by Canada's supposedly wonderful health and social support systems; the very systems that let Ruth down.[76] Changing these systems can only happen when equality rather than cost-effectiveness, fairness rather than finance, and compassion

---

[71] See Chapter 1, The Rotor.
[72] See Chapter 1, The Rotor.
[73] "Calcedonies," as published in the *Canadian Medical Association Journal* (Nisker, 2001a).
[74] Nisker, 2012.
[75] Nisker, 2015b.
[76] Nisker, 2010.

rather than counting become the prevailing imperatives in Canadian healthcare.

Victor was a gentle giant, the strongest and kindest man I have ever met. Victor did not complain of the growing pain lancinating his core until family members observed that his skin had turned yellow. Victor finally agreed to see a doctor, and eventually endured massive surgery for pancreatic cancer. However, the cancer was more "aggressive" than the surgeon had anticipated. Victor's strength permitted rapid recovery to soon resume his daily rounds of assisting family members and friends, until four months later cancer recurred. Victor's strength then endured the wrath of pancreatic cancer's advance.

While I was writing Victor's story five years after his death, and more than five years after the pancreatic-cancer death of two of his sisters and his brother, one of Victor's nieces was dying of pancreatic cancer. She is the exact age of Victor's daughter Roxanne. I am consumed with worry for Roxanne, and with the injustice of the reluctance to fund research on hereditary pancreatic cancer. The reason for this reluctance is that hereditary pancreatic cancer is a rare form of pancreatic cancer, and pancreatic cancer is a rare form of cancer. However, pancreatic cancer is the second most common cancer-killer.

Canadian COVID Injustice on Beaches and Beach Volleyball Courts attempts to capture my concern regarding the refusal of "social distancing" of twenty-somethings in Canada. Although I begin with this refusal on beaches and beach volleyball courts in 2020 and 2021, both of which "spiked"[77] COVID-19 in our university town,

---

[77] "Spike" is a term used in volleyball for hitting the ball hard over the net, and sharply down on the other side.

along with the "spikes" induced by parades of students cheering "freedom"[78] through campus and surrounding streets during "Frosh Week"[79] and St. Paddy's Days.[80]

This chapter flashes back 45 years to Toronto's Sunnyside Beach, where my Grandmother social distanced me from other children. The virus my Grandmother feared is much less contagious than the COVID virus, especially the Omicron variant spreading rapidly in 2022, but the polio virus generally has more serious consequences for children. Dr. Salk's vaccine made him my childhood hero, and he became my medical-school hero when I learned he had refused to patent his vaccine. The reason for his refusal concluded with, "Could you patent the sun?"[81] I had the gift of visiting the Salk Institute in La Jolla, California, and feeling the inspiration of the man whose vaccine did so much for humanity. As I walked the beaches between the Salk Institute and the sea, I had no idea these beaches would soon spread COVID-19.

---

[78] Anti-masker marches in Canada; "Freedom" has been the chant of opposition to social distancing since the first COVID lockdown in March 2020. "Freedom" became the mantra for the "antivaxxers" (Hotez, 2021; Kurjata, 2021), who represented over 70 percent of persons in our ICUs in January 2022 (Daigle, 2022; Favaro & Jones, 2022), and who continue to be an open invitation to new COVID variants. "Freedom" later became the mantra for the antivaxxer truckers who convoyed across Canada to Parliament Hill in Ottawa to protect against our vaccine mandate ("COVID-19 protesters demonstrate across Canada in support of truck convoy in Ottawa," 2022; Andrews & Anand, 2022).

[79] Frosh Week is formally referred to as "Orientation Week" by most universities.

[80] Our students are but a microcosm of the many self-centred Canadian decision-makers who privileged personal "freedom" over community responsibility during COVID including antivaxxers, whether they were in air-horn blaring "freedom"-convoys of antivaxxer truckers in January and February 2022, or quieter Canadians with similar antivaxxer views.

[81] Bos, 2013.

Webinar Physicians' Cavalier Terms for COVID-Ventilator Triage of Disabled Persons[82] expresses my outrage at the nonchalant terms used by Canadian critical-care physicians on a COVID webinar in 2021, insisting restriction from COVID ventilators[83] of persons deemed "less healthy."[84] Their insistence was based on the myth of "finite health resources"; a lack-of-social-justice myth that sentences socio-economically disadvantaged persons to inadequate healthcare. The cavalier terms used by these critical-care physicians were heard by over 300 healthcare professionals, and likely spread exponentially through subsequent online seminars and presentations at conferences.

It is imperative to disseminate the contrary view: that a small increase in taxes, rather than restriction from ventilators, is the ethical strategy to ensure social justice for socio-economically disadvantaged persons in times of COVID. Physicians must never be judges and juries with the capacity to triage persons from COVID ventilators.

COVID Injustice Before I Heard the Word "COVID" recounts early March 2020, when I had not yet learned of the "virus of concern," even though it had been spreading in China since autumn 2019, even though I am a physician, even though I developed pneumonia from this virus. Canadian winter had encouraged my partner and I to flee to Mexico for our University's Reading Week;[85] however, on the third day, fever, coughing, and difficulty breathing led me to the doctor who diagnosed left-lower lobar pneumonia. My partner found a seat for me on the next flight back to Canada. Although my difficulty breathing worsened in the jet and taxi home,

---

[82] Nisker, 2021b.
[83] Wilson, 2021.
[84] Dubinsky, McKenna, Loiero, & Leung, 2021.
[85] "Reading Week" has become the ironic misnomer for the fun-break in February that universities afford students (and professors), presumably to catch up on their work.

my fear of being put on a ventilator inhibited me from asking the cab driver to stop at a hospital.

In this chapter I contend that the Chinese government's suppression of knowledge of the virus that would later be called COVID-19 was the greatest injustice in medicine perpetrated in my lifetime. As I am finishing this chapter in May 2022, there have been 40,843 COVID-19 deaths reported in Canada,[86] and at least 6,283,594 deaths worldwide.[87] This injustice must never happen again, but will happen again unless opaque governments are transparent regarding the spread of viruses.[88]

COVID Aggression Condemns a Muslim Family Near Our Medical School mourns the tragedy a few blocks from my home on June 6, 2021, when a right-wing extremist ploughed his truck into a Muslim family taking their evening walk. As soon as I learned about this tragedy, I pilgrimaged to its site, placed flowers from my garden on the growing mountain of flowers, and knelt with others absorbing the injustice that had occurred. The following day as I entered the Hospital where my office is located, I felt a vacuum of compassion for the nine-year-old boy here, suffering fractures and organ injuries, inflicted by the hate-inspired driver that murdered his parents, his sister, and his grandmother. Who would tell the boy what had happened to his family?

In this chapter I worry that even in a liberal democracy like Canada, KKK-type white-supremacist hate is alive and well beneath the surface and is ready to rear its ugly head of aggression, as happened in this attack on a family for the sin of being Muslim.

---

[86] John Hopkins University of Medicine, 2022.
[87] John Hopkins University of Medicine, 2022.
[88] The Omicron variant was the most contagious strain of COVID-19, and became the dominant strain by the beginning of 2022 (Khandia, 2022).

The Lottery conveys my shock while listening to the Canadian Broadcasting Corporation News on January 19, 2022, informing Canadians that physicians in Toronto were considering a "lottery" system for determining which patients would have admission to intensive care as "critical medicines" were "in short supply."[89] My nausea with the concept of a lottery for admission to intensive care was amplified by the echo of Shirley Jackson's short story, "The Lottery", written in 1948.[90] In Jackson's story, the residents of a late 1940s American town draw paper slips happily from a metal box, analogous to the way many people pick up lottery tickets from supermarkets in 2022. The anticipation of "The Lottery" in Jackson's story was exciting, even jovial, and the day of the lottery seemed to be a picnic-like celebration. However, the reader gradually senses darkness in "The Lottery," and eventually realizes that the winner of "The Lottery" will receive a prize the reader would rather not receive.

In this chapter, I juxtapose Shirley Jackson's lottery with Canadian critical-care physicians' recommendation in 2022 that ventilators by lottery should not be ruled out, as lottery is the "the least unjust form of allocation in the coronavirus disease."[91] I rebut their recommendation with the point that Canada ought to always ensure justice in healthcare, rather than ever pursuing "the least unjust form," even if it means an increase in taxes to purchase more ventilators and fund the personnel to operate them.

In **Antivaxxer Xenophobic COVID Violence**, I describe the first months of 2022, when antivaxxers overflowed Canada's ICUs, while other antivaxxers condemned the vaccine mandates that were put in place to keep people out of those ICUs. Caravans of antivaxxer truckers blatantly blared their airhorns across Canada to Ottawa and blocked border-crossings to the United States at Windsor, Ontario,

---

[89] Canadian Broadcasting Corporation News January 19, 2022.
[90] Jackson, 1948.
[91] Silva, 2020.

Emerson, Manitoba, and Coutts, Alberta. These white antivaxxer truckers wore Nazi swastikas on their arms and waved Confederate flags reminiscent of black slavery's past, and the Ku Klux Klan's past and present.

The antivaxxer truckers' words and actions were venomous, but worse was the aggressive sympathy they generated in a small, but potentially dangerous, percentage of white Canadians that could result in more of the violence that condemned the Muslim family near my home in Chapter 22. The aggression against the Muslim family was followed by four sexual assaults in our University's residences, as well as a manslaughter on campus. I contend these copy-cat aggressions have no place in Canada, nor in any country, regardless of the animus generated by COVID.

The Arrogance of "But All You Need Is a Good Index Finger" explores the injustice of lack of public funding of screening for prostate cancer, the most common cancer in men over 50. This narrative is a happy story, though it should not have been, as I have aggressive metastatic prostate cancer. Perhaps my cancer is metastatic because I routinely refused my family physician's recommendation to have annual PSA[92] testing; refused because PSA testing was not publicly funded. After all, I teach ethics and social justice to medical students, and must walk the talk. I may have also refused PSA testing because of the certainty of the words of a Canadian physician at a prestigious national conference, claiming "But all you need is a good index finger."

My PSA story is a happy story only because I am a physician-educator at a major medical school and so experienced a string of exceptionalisms and coincidences that have allowed me to still be alive to write my happy PSA story. Indeed, I am compelled to write this story as a *mea culpa* to the many men similarly situated to me who have died

---

[92] PSA is the short form for "Prostate Specific Antigen."

from prostate cancer, sacrificed by statisticians to cost-effectiveness imperatives.[93] My PSA story is a happy story, and will remain a happy story when my life comes to its inevitable termination from prostate cancer. Initiation of public funding of annual PSA testing for all men over 50 would make my PSA story an even happier one.

Our Third COVID Summer explores the inconsistencies between Public Health proclamations of concern and our governments' political ambitions that cause them to lift restrictions. Although COVID is still with us, albeit in its usual summer dip, Public Health's prediction of an explosion in the autumn of Omicron variants causes physicians, including me, to be very concerned.

In "Our Third COVID Summer," I not only reflect on the discrepancy between "open for business" governments and Public Health caution, but describe the frustrations leading to the aggression of ordinary Canadians. This aggression is seen in signs in bank windows pleading "abuse of tellers will not be tolerated" and in the aggression of car drivers, whether cutting in and out of traffic or pushing joggers like me off the road. This particular aggression is usually accompanied by a certain finger.

In "Our Third COVID Summer," I experience personal conflict regarding being the only person wearing a mask because of my low immunity, whether I should be the only person wearing a mask, and the effect of my continued mask-wearing on those around me, including my grandchildren. My conflict also arises from the fact that even with triple COVID vaccination, many people suffer COVID, albeit perhaps not as severely as if they were not vaccinated. After all, when triple-vaxxed prominent persons, such as prime ministers and presidents, still come down with COVID, why should ordinary citizens believe that they should be vaccinated?

---

[93] Bell, Connor Gorber, Shane, Joffres, Singh, *et al.*, 2014; Nisker, 2020a; Prostate Cancer Canada, 2017.

References

(2022, Jan 29). COVID-19 protesters demonstrate across Canada in support of truck convoy in Ottawa. Canadian Broadcasting Corporation News. https://www.cbc.ca/news/canada/canada-protests-truck-convy-1.6332680

Andrews B & Anand A. (2022, Jan 31) After weekend of protests, Ottawa residents are feeling the effects. Canadian Broadcasting Corporation News. https://www.cbc.ca/news/canada/ottawa/convoy-workers-two-days-later-1.6333017

Azpiri, J. (2021, May 28). *Rallies against B.C. old-growth logging held at offices of premier, attorney general. Global News.* https://globalnews.ca/news/7901871/bc-fairy-creek-old-growth-logging-protest-horgan-eby/

Ballingall, A. (2016, Feb 10). *Former Huronia residents join speakers' series to educate others on horrors endured.* Toronto Star. https://www.thestar.com/news/gta/2016/02/10/former-huronia-residents-join-speakers-series-to-educate-others-on-horrors-endured.html

Bell, N., Connor Gorber, S., Shane, A., Joffres, M., Singh, H., Dickinson, J., Shaw, E., Dunfield, L., Tonelli, M., & Canadian Task Force on Preventive Health Care. (2014). Recommendations on screening for prostate cancer with the prostate-specific antigen test. *Canadian Medical Association Journal, 186*(16), 1225–1234.

Bos, C. (2013). *Jonas Salk—"Could you patent the sun?"* Awesome Stories. https://www.awesomestories.com/asset/view/Jonas-Salk-Could-You-Patent-the-Sun-1

Burlock, A. (2017). *Women with disabilities.* Statistics Canada. https://www150.statcan.gc.ca/n1/pub/89-503-x/2015001/article/14695-eng.htm

Cambie Surgeries Corp. v. British Columbia (Medical Services Commission), 2010 BCCA 396. Court of Appeal for British Columbia, 2010.

Cribb, R., & Isai, V. (2017, Oct 17). *Should the wealthy be allowed to buy their way to faster health care at private clinics?* Toronto Star. https://www.thestar.com/news/canada/2017/03/18/should-the-wealthy-be-allowed-to-buy-their-way-to-faster-care-at-private-clinics.html

Daigle, T. (2022, Jan 15). In this Ontario hospital, it's mostly the unvaccinated who are overwhelming the ICU. Canadian Broadcasting Corporation News. https://www.cbc.ca/news/health/sarnia-bluewater-health-hospital-covid-patients-1.6315681

de Saint-Exupéry, A. (1943). *The Little Prince.* Harcourt Brace Jovanovich.

Dubinsky, A., McKenna, T., Loiero, J., & Leung, A. (2021, Apr 19). *As ICUs fill up, doctors confront grim choice of who gets life-saving care.* Canadian Broadcasting Corporation News. https://www.cbc.ca/news/health/covid-ontario-icu-triage-1.5992188

Favaro A & Jones A. (2022, Jan 12). Inside an ICU where 70 per cent of COVID-19 patients are unvaccinated. CTV News. https://www.ctvnews.ca/health/coronavirus/inside-an-icu-where-70-per-cent-of-covid-19-patients-are-unvaccinated-1.5738198

Frazee, C., Gilmour, J., Mykitiuk, R., & Bach, M. (2002). The legal regulation and construction of the gendered body and of disability in Canadian health law and policy. *National Network on Environments and Women's Health.* http://www.cwhn.ca/en/node/24745

Gaetz, S., Donaldson, J., Richter, T., & Gulliver, T. (2013). *The state of homelessness in Canada 2013.* Canadian Homelessness Research Network Press. http://homelesshub.ca/sites/default/files/SOHC2103.pdf

Gordon, W. A. (1990). *The fourth of May: Killings and coverups at Kent State.* Prometheus Books.

Government of Canada. (2004). *Assisted Human Reproduction Act* (S.C. 2004, c. 2). https://laws-lois.justice.gc.ca/eng/acts/A-13.4/page-1.html#h-6052

Government of Canada. (2017). *Genetic Non-Discrimination Act* (S.C. 2017, c. 3). https://laws-lois.justice.gc.ca/eng/annualstatutes/2017_3/

History.com. (2010). *Martin Luther King, Jr. delivers "I have a dream" speech at the March on Washington*. This Day in History. A&E Television Networks. https://www.history.com/this-day-in-history/king-speaks-to-march-on-washington

Hotez, P. (2021). COVID vaccines: time to confront anti-vax aggression. Nature, 592(7856), 661. https://doi.org/10.1038/d41586-021-01084-x

Jackson S. (1948). The Lottery. The New Yorker.

John Hopkins University of Medicine. (2022, Feb 16). COVID-19 Dashboard https://coronavirus.jhu.edu/map.html

Jolie, A. (2013, May 14). *My medical choice*. New York Times. https://www.nytimes.com/2013/05/14/opinion/my-medical-choice.html

Joseph, M., Rab, F., Panabaker, K., & Nisker, J. (2015). Feelings of women with strong family histories who subsequent to their breast cancer diagnosis tested BRCA positive. *International Journal of Gynecological Cancer* 25(4), 584–592.

Khandia R, Singhal S, Alqahtani T, Kamal MA, El-Shall NA, Nainu F, Desingu PA, Dhama K. (2022). Emergence of SARS-CoV-2 Omicron (B.1.1.529) variant, salient features, high global health concerns and strategies to counter it amid ongoing COVID-19 pandemic. Environ Res. doi: 10.1016/j.envres.2022.112816. Epub ahead of print. PMID: 35093310; PMCID: PMC8798788.

Kurjata, A. (2021, Oct 22). Husband regrets anti-vaxx stance as wife lies in a coma 800 km from home. Canadian Broadcasting Corporation News. https://www.cbc.ca/news/canada/british-columbia/anti-vaccine-fort-st-john-pregnant-wife-1.6222325

Mikkonen, J., & Raphael, D. (2010). *Social determinants of health: The Canadian facts*. York University School of Health Policy and Mangement. http://thecanadianfacts.org/The_Canadian_Facts.pdf

Mykitiuk, R., & Nisker, J. (2010). Social determinants of "health" of embryos. In J. Nisker, F. Baylis, I. Karpin, C. McLeod, & R. Mykitiuk (Eds.), *The "healthy" embryo: Social, biomedical, legal and philosophical perspectives* (pp. 116–135). Cambridge University Press.

Narod, S. A., Brunet, J. S., Ghadirian, P., Robson, M., Heimdal, K., Neuhausen, S. L., Stoppa-Lyonnet, D., Lerman, C., Pasini, B., de los Rios, P., Weber, B., Lynch, H., & Hereditary Breast Cancer Clinical Study Group. (2000). Tamoxifen and risk of contralateral breast cancer in BRCA1 and BRCA2 mutation carriers: A case-control study. Hereditary Breast Cancer Clinical Study Group. *Lancet, 356*(9245), 1876–1881.

Narod, S. A., Feunteun, J., Lynch, H. T., Watson, P., Conway, T., Lynch, J., & Lenoir, G. M. (1991). Familial breast-ovarian cancer locus on chromosome 17q12-q23. *Lancet, 338*(8759), 82–83.

Narod, S., Lynch, H., Conway, T., Watson, P., Feunteun, J., & Lenoir, G. (1993). Increasing incidence of breast cancer in family with BRCA1 mutation. *Lancet, 341*(8852), 1101–1102.

Nisker J, Bergum V. (2007). A Child on Her Mind. In: Bergum V, Van Der Zalm J (Eds.). *Mother Life Studies of Mothering Experience.* Edmonton (Canada): Pedagon Publishing, p.364-398.

Nisker, J. (1995). Beneath the pagoda's perch: A Canadian at Kent State. In *Poetry about Kent State Shootings.* Kent State Shootings: Digital Archive. https://omeka.library.kent.edu/special-collections/items/show/4448

Nisker, J. (2001a). Chalcedonies. *Canadian Medical Association Journal, 164*(1), 74–75.

Nisker, J. (2001b). Orchids: Not necessarily a gospel. In J. Murray (Ed.), *Mappa Mundi: Mapping Culture/Mapping the World* (pp. 61–110). University of Windsor Press. http://www.uwindsor.ca/hrg/mappa-mundi-mapping-culturemapping-the-world-table-of-contents-0

Nisker, J. (2010). Calcedonies: Critical reflections on writing plays to engage citizens in health and social policy development. *Reflective Practice, 11*(4), 417–432.

Nisker, J. (2012). *From Calcedonies to Orchids: Plays promoting humanity in health policy.* Iguana Books.

Nisker, J. (2013). A public health education initiative for women with a family history of breast/ovarian cancer: Why did it take Angelina Jolie? *Journal of Obstetrics and Gynaecology Canada, 35*(8), 689–691.

Nisker, J. (2015a). The latest thorn by any other name: Germ-line nuclear transfer in the name of "mitochondrial replacement." *Journal of Obstetrics and Gynaecology Canada, 37*(9), 829–831.

Nisker, J. (2015b). *Patiently waiting for...* Iguana Books.

Nisker, J. (2018). The cement spiral. *Journal of Obstetrics and Gynaecology Canada, 40*(6), 643–645.

Nisker, J. (2019). Dissolution of Canada's single-tiered health system would threaten the health of women with disabilities. *Journal of Obstetrics and Gynaecology Canada, 41*(11), 1616–1618.

Nisker, J. (2020). Arrogance of "but all you need is a good index finger": A narrative ethics exploration of lack of universal funding of PSA screening in Canada. *Journal of Medical Ethics, 46*(4), 249–252.

Nisker, J. (2021). A brief and personal history of "what's in a name" in reproductive genetics. *Medical Humanities, 47*(2), 228–234.

Nisker, J. (2021, May 13). *Webinar physicians' cavalier terms regarding triage from COVID ventilators.* Impact Ethics. https://impactethics.ca/2021/05/13/webinar-physicians-cavalier-terms-about-triage-from-covid-ventilators/

Nisker, J., Baylis, F., Karpin, I., McLeod, C., & Mykitiuk, R. (Eds.). (2010). *The "healthy" embryo: Social, biomedical, legal and philosophical perspectives.* Cambridge University Press.

Nisker, J. (2022). Confined to the COVID Sidelines. In *Confined to the COVID Sidelines New and Selected Verses in the Time of COVID.* In press.

Nisker, J.A. (2007). The need for public education: "Surveillance and risk reduction strategies" for women at risk for carrying BRCA gene mutations. *Journal of Obstetrics and Gynaecology Canada*, *29*(6), 510–511.

Nisker, J.A., & Gore-Langton, R. E. (1995). Pre-implantation genetic diagnosis: A model of progress and concern. *Journal of Obstetrics and Gynaecology Canada*, *17*(3), 247–262.

Paikin, S. (2016). *Bill Davis: Nation builder, and not so bland after all*. Dundurn.

Preibisch, K., & Hennebry, J. (2011). Temporary migration, chronic effects: The health of international migrant workers in Canada. *Canadian Medical Association Journal*, *183*(9), 1033–1038.

Prostate Cancer Canada. (2017). *Statistics*. http://www.prostatecancer.ca/Prostate-Cancer/About-Prostate-Cancer/Statistics

Royal College of Physicians and Surgeons of Canada. (2021). *CanMEDS: Better standards, better physicians, better care*. http://www.royalcollege.ca/rcsite/canmeds/canmeds-framework-e

Shakespeare, W. (1734). *Romeo and Juliet. By Mr. William Shakespeare*. Eighteenth Century Collections Online. Gale. https://www.gale.com/primary-sources/eighteenth-century-collections-online

Shuchman, M. (2020). Theatre of social justice. *Canadian Medical Association Journal*, *192*(23), E636–E637.

Siegler, M., & Epstein, R. A. (2003). Organizers' introduction to the Symposium on Quality Health Care. *Perspectives in Biology and Medicine*, *46*(1), 1–4.

Silva D. S. (2020). Ventilators by Lottery: The Least Unjust Form of Allocation in the Coronavirus Disease 2019 Pandemic. Chest, 158(3), 890–891. https://doi.org/10.1016/j.chest.2020.04.049

Silversides, A. (2008). Canada Health Act breaches are being ignored, pro-medicare groups charge. *Canadian Medical Association Journal*, *179*(11), 1112–1113.

Smylie, J., Firestone, M., Spiller, M. W., & Tungasuvvingat Inuit. (2018). Our health counts: Population-based measures of urban Inuit health determinants, health status, and health care access. *Canadian Journal of Public Health, 109*(5–6), 662–670.

Statistics Canada. (2017). *Health fact sheets: Primary health care providers, 2016.* https://www150.statcan.gc.ca/n1/pub/82-625-x/2017001/article/54863-eng.htm

University Health Network. (2021). Princess Margaret History. https://www.uhn.ca/corporate/AboutUHN/OurHistory/Pages/princess_margaret_history.aspx

Vanstone, M., Cernat, A., Nisker, J., & Schwartz, L. (2018). Women's perspectives on the ethical implications of non-invasive prenatal testing: A qualitative analysis to inform health policy decisions. *BMC Medical Ethics, 19*, article 27.

Vanstone, M., King, C., de Vrijer, B., & Nisker, J. (2014). Non-invasive prenatal testing: Ethics and policy considerations. *Journal of Obstetrics and Gynaecology Canada, 36*(6), 515–526.

Vanstone, M., Yacoub, K., Winsor, S., Giacomini, M., & Nisker, J. (2015). What is "NIPT"? Divergent characterizations of non-invasive prenatal testing strategies. *AJOB Empirical Bioethics, 6*(1): 54–67.

Wilson, C. (2021, Jan 18). *Ontario patients to be ranked for life-saving care should ICUs become full, documents show.* CTV News. https://toronto.ctvnews.ca/ontario-patients-to-be-ranked-for-life-saving-care-should-icus-become-full-documents-show-1.5271774

World Bank Group. (2018). *Physicians (per 1,000 people).* World Bank data. https://data.worldbank.org/indicator/SH.MED.PHYS.ZS?end=2015&locations=CA&start=1961&view=char

World Health Organization. (2010). *A conceptual framework for action in the social determinants of health.* World Health Organization. https://www.who.int/publications/i/item/9789241500852

Chapter 1

The Rotor

They think I have MS. My family doctor, the neurologist she referred me to, the subspecialist neurologist he referred me to, the neuro-ophthalmologist she referred me to all think I have MS. They speak to me in noncommittal medical euphemisms rather than "burdening" me with a diagnosis that is not yet "definitive," but they think I have MS. I welcome their avoidance of a not-yet "definitive" diagnosis that will be so definitive for me.

I have developed an intermittent paralysis of the small muscles that control the movement of my right eyeball. Small muscles, like eye muscles, I understand, may weaken first with MS. Two or three times a day, in ten- to fifteen-minute intervals, the walls spin because my right eyeball is "rotating out of control" like an out-of-control handheld camera, and I find myself plastered against the nearest wall. At least, I am told that my right eyeball is "rotating out of control," never having been able to observe this phenomenon myself, as I cannot see anything during my right eyeball's delinquency. Eventually I learn to cover up my right eye so I can see through my still well-behaved left eye, albeit not stereoscopically, but my left eye cannot see my right eyeball rotate because my right eye is covered up by my right hand.

Covering up my right eye allows me to ignore what is going on. Covering up my right eye allows me to cover up from my clinical colleagues what is going on in someone on whom they rely and require to function at 100 percent capacity now and in the future. Although none of my colleagues ever comments, those who observe me standing tipsily or intermittently wobbling along the Hospital's walls from our offices to the Clinic with my right hand covering my right eye know something is going on, but politely ignore that something is going on. I am grateful for their feigned lack of observation. Covering up my right eye also allows me to cover up what is going on from my patients. Although the occasional woman who witnesses me struggling to look in her eyes with my left eye while scrawling her presenting symptoms and history with my untrained left hand on her chart knows something is going on, I never see mistrust in their eyes. That would be too hard.

Of course, I am sent for an MRI. For those of you who have not had the privilege, you deposit your clothes, wallet, and "any loose metal" (as opposed to metal in a replacement joint, cardiac pacemaker, or intrauterine contraceptive device) in a locker outside the magnet's room. Then, in a standard open-backed, greyish-blue hospital gown, you lie on a greyish-white slab that feels like cold steel, but must be plastic or fibreglass, considering the slab with you on it will be rolled into the mouth of a huge and powerful magnet. A reassuring technician Velcro-straps your head down to the slab and places a panic button in your right hand, "Just in case." As the slab slowly rolls in, you feel like you are in a drawer being rolled into the storage cabinet wall of a morgue. The hair on your arms brushes the narrow sides of its walls. Your eyes stare at the seeming-to-move lid a few centimetres from your face. If you do not possess the good sense to close your eyes, which I do not, claustrophobic waves may wash over you, and may begin to drown you, and your thumb may caress the panic button with increasing frequency, and you may push the panic button, and hear the technician's trying-to-calm-you voice as she rolls you out of the magnet, and tells you to "Take some deep breaths" and "Think pleasant thoughts" and "Try closing your eyes next time" before she rolls you back in.

While the magnet rotates around me and my tightly closed eyes, I try to distract claustrophobia's imminent implosion with "pleasant thoughts" of running free in my childhood. The whirling clicks and grinds of the magnet's rotation bring me to a midway ride at the Canadian National Exhibition called The Rotor.

My friends and I pilgrimaged to "The Ex" every year on Labour Day weekend, as did tens of thousands of other kids. The guys met at my house early in the morning. We walked the block to catch the TTC bus, then eagerly transferred to a streetcar in "the loop," then another streetcar, without parental accompaniment since age ten, the year of that first Rotor ride. Our intense anticipation inflated with each clang of the streetcar's overhead cord, pulled by a rider requesting the next stop. We could reach the cord only by standing on a seat, which we only occasionally did, just before we accidentally pulled the cord. Each announcement of the final streets' names bringing us closer to and then through The Ex's massive Dufferin Gate was met with exhilarated cheers.

After wildly whooping as we flew off the streetcar, we would run as straight as possible to a roller coaster. Unlike my friends, I did not love roller coasters; quite the opposite, the drops terrified me. However, The Rotor terrified my friends, while, for some reason, I found it peaceful. And as I always had to appear courageous to my friends of course, The Rotor became an important part of my childhood.

The Rotor is a round metal room about 50 feet in diameter, covered by a conical tin roof. There are no restraining bars or shoulder harnesses in The Rotor, like those compulsory in the other high-speed rides. Indeed, there are no leather straps or plastic handles, or indeed anything to clutch onto in The Rotor. You just stand against the 360° wall and stare across its vacant diameter at the person standing against the opposite wall.

Soon you hear an ominous metallic thud as the small steel door through which you entered, pretending to duck your head while still standing on your tiptoes to meet the height requirement, is shut. Then you hear the scary scrape of rusty steel sliding on rusty steel as the door is bolted shut from the outside. After a second or two of hermetically sealed silence, you hear the molten groan and feel the grating jerk of The Rotor rousing from sleep.

As The Rotor commences a slow rotation, you widen your stance on the corrugated steel floor, bend your knees a bit, spread your arms, and lean harder and harder into the wall behind you. You hear the clacketing of steel wheels increase in frequency and pitch, as they accelerate on what must be a circular steel track below the floor. Soon you no longer have to lean into the wall behind you, as your back, then head, then arms have become plastered onto it. You cannot move anything except your eye muscles. Then, to your shock on your first time on the ride the floor drops out and your eyeballs, which you try hard to restrain from looking down, stupidly stare into the screaming jaws of the beast below your unsupported feet. You panic, as you are sure that at any second you will slide down the wall to be shredded by its cogs and gears. You mentally try to suction yourself against the wall behind you because you physically can't use the muscles you need to press your back into it. As you rotate faster and faster, you stop worrying about sliding down the wall and start worrying that you are going to take off upward like a rocket and crash your head into the metal roof. However, if you are very lucky, you may relax and bask in the peaceful elation of gravity defied.

Too soon The Rotor begins to slow, and you actually do start sliding down the wall, and start clamouring with all your now-working muscles to cling to the wall to avoid being chewed by The Rotor's gears. You slide down anyway, but the floor is miraculously back. You gratefully embrace the solidity of its steel, preferably with your feet, but sometimes with your butt or a shoulder. You hear the bolt outside the door scrape again and clang open like on a submarine after surfacing. You follow the curved line of tipsy Rotor riders and wobble along the wall to the door, trying not to bump into the rider ahead of or behind you. You wobble through the door to the concrete pavement outside, past a man with a mop and bucket and dangling cigarette, to join a chorus of vomiters, gushing out the corn dogs and cotton candy gorged since their arrival at The Ex.

I rode The Rotor many times in the subsequent years of my youth, but always taking a two-hour food pause before ducking through its door, and always wishing the other riders did the same, particularly those who vomited while The Rotor was still in spin. My friends never rode The Rotor again. They roller-coastered while I rotated, personally preferring centrifugal force to gravitational force. When we met at a prescribed time and cotton candy location, they would always shake their heads with a mixture of awe and concern for my sanity when they saw I had once again survived The Rotor intact—that is, with no vomit on my T-shirt. One year The Rotor was no longer there. The midway had become too frivolous for our times anyway.

Lying in the magnet, eyes smiling but firmly shut, brain firmly fixed in my youth, I reach peace, and begin investigating the dichotomy of rotating in The Rotor, centrifugally plastered vertically on its metal wall, and having a magnet rotating around me, gravity and Velcro tape holding me down horizontally to its non-metal slab. I may have fallen asleep, which would not have been unreasonable, considering the MRI began at 0200 hours. The next thing I hear is, "It's over," and I feel the drawer being rolled out.

"That wasn't so bad, was it?" the technician gently soothes, as if I am a child or an elderly person.

I respond, "Next time I want a blindfold like they give you in front of a firing squad."

It is a bad joke, and I am certain I will be punished.

I am called down to see the neurologist the next day to hear the result. "Normal" would have come over the phone. The MRI showed white specks at the base of my brain and in the brain's balance centre, the cerebellum. These white specks could mean MS. My neurologist remains noncommittal as to their meaning. I am grateful that he continues to avoid articulating a possible diagnosis.

The episodes of my right eyeball's rotating increase in frequency and duration, prompting a second MRI only three months after my first. The next day, I go over the computer images with one of the neuroradiologists. He stares at the images, lifts his head, and apologizes when he tells me that "The number of white specks has significantly increased." He looks at my brain's images, then my eyes, back to my brain, and finally settles on my eyes.

"What's going on?" slips from my mouth before I can zip it shut. I can't put it back.

His eyes go back to the MRI images. "I'm sorry, I'm not sure what is going on."

Later in the day, the neurologist echoes, "Not sure."

Two nights later, I bolt upright from sleep with a tremendous headache. I do not get headaches. I am surprised to see a torrent of vomit hurl against the wall on the opposite side of the bedroom. Vomit hits the wall again and again, "projectile vomiting" like Linda Blair's in *The Exorcist*. I have not vomited since my first Rotor ride.

So, I know what is going on as I am being driven at full speed to the Emergency Room, vomit bursts slamming the bottom of my bathroom's beige plastic wastebasket. The ER nurses know what is going on as they hurry my stretcher to the MRI suite, after quickly handing me a kidney-shaped aquamarine plastic vomit basin. The MRI technician knows what is going on as she quickly Velcros down my forehead and rolls me into the magnet, telling me, "Try not to vomit." And when she rolls me out of the magnet and I see the Chief of Neurosurgery, the Chief of Neuroradiology, the neurologist-on-call, and my neurologist wringing eight hands in front of a wall's worth of my brain's computerized images, I definitely know what is going on.

The doctors step toward me in unison, a tight phalanx of sombre centurions. The smile on my face surprises them and makes them feel uncomfortable. The Chief of Neuroradiology casts down his eyes and reaches out his right hand to my left shoulder.

"Jeff, you've had an event."

When referring to one's brain, "event" is the euphemism for a stroke, usually a thrombotic stroke, as "a bleed" is the euphemism for a hemorrhagic stroke. I test my brain's speech centre with, "Yes, I know." My smile broadens. Their expressions become more grave.

"Do — you — un — der — stand what I mean?" the Chief of Neurology slowly asks, drawing out and emphasizing each syllable.

I laugh, "Yes, I know exactly what you mean."

Then I pump my first "Yes." I test my legs by sliding off the slab and standing. I pump my fist, shout "Yes" again; then raise my arms in Victory's V. I proceed to take a wobbly victory lap around the MRI room, bouncing off the occasional wall. When I return to the doctors, I try to give each a high five. But none of them is interested in high-fiving a once-serious colleague who has clearly been rendered both mentally and physically unbalanced by a stroke.

Their embarrassed eyes surreptitiously flit to one another and then back to the deranged dervish. Most eyes eventually settle on my neurologist. He knows me best. He slowly steps forward, lost in thought. I assume he's contemplating his neuroanatomy knowledge to determine whether the location of my stroke could have, in addition to obviously affecting my brain's balance centre, also affected its mood centre. He puts his right hand on my left shoulder, and says in a very compassionate voice, "Do you understand what we are telling you, Jeff?"

"Yes."

"What do you understand?"

"That I don't have MS."

Another version of "The Rotor" was previously published in *Patiently Waiting For...*[94]

---

[94] Nisker, 2015.

References

Nisker, J. (2015). *Patiently waiting for...* Iguana Books.

Chapter 2

"You Must Go to Medical School or Hitler Will Have Won"

It has been a gift being a physician, though being a physician was not what I had planned. I had been groomed from birth by my Grandmother to be a human rights lawyer, to go further in the pursuit of social justice than her lack of education and young death from breast cancer permitted her.[95] I had also been encouraged to be a human rights lawyer by each of my high school English teachers because I always chose a social-justice topic for public speaking competitions. I had been inspired to be a human rights lawyer by Atticus Finch, the lawyer in *To Kill a Mockingbird*,[96] who defended a wrongly accused Black man amidst segregation and hatred in the 1930s' American South. I had also been inspired to be a human rights lawyer by Dr. Martin Luther King Jr.,[97] to

---

[95] See Chapter 5, Princess Margaret; See Chapter 10, She Lived with the Knowledge; See Chapter 13, The "Helix of Life" Revisited: DNA in Concrete and Not.
[96] Lee, 1960.
[97] The novel *To Kill a Mockingbird* was written by Harper Lee (1960). Atticus Finch was portrayed by Gregory Peck, who won an Academy Award in the film version directed by Robert Mulligan (1962).

see his dreams of equality become reality in my generation. It has been a gift being a physician.

As I walked home from basketball practice on a cold and dark January evening, I shuddered upon seeing my Uncle's old black car in our driveway. My Uncle never visited us on weeknights. His car in our driveway was sinister, as it reminisced my Grandfather's car in our driveway as I walked home from practice a year ago, the night I was told my Grandmother had just died.[98] I ran the last two blocks home.

On opening the front door, I saw my Uncle sitting at the head of our "kitchenette" table. Our "kitchenette" consisted of a whitish "leatherette" horseshoeing a Formica-topped, chrome-legged square table and a matching chair at the horseshoe's opening. Copperish thumbtacks studded the perimeters of the leatherette and chair at half-inch intervals. My Father was sitting on the leatherette's left wing, invisible as usual behind his *Toronto Star*. My Mother was sitting on the leatherette's right wing, staring down at a pale-green plastic tablecloth. The empty chair beckoned me. This tableau was set by something unfortunate.

On a usual evening, by the time I came home from whichever practice was in season, my Father would have already finished dinner and be religiously fixed in front of Walter Cronkite.[99] On a usual evening, my aproned Mother would be religiously fixed at the kitchen sink doing the dishes. On a usual evening, I would have snuck up behind my protesting Mother and hugged her while she was vulnerable to such hugs because of soapy rubber-gloved hands. On a usual evening, I would have taken my "keeping it warm" dinner plate

---

[98] See Chapter 5, Princess Margaret; See Chapter 10, She Lived with the Knowledge; See Chapter 13, The "Helix of Life" Revisited: DNA in Concrete and Not.

[99] Walter Cronkite was the well-respected CBS Evening News anchor from 1962 to 1981. In 1963, it was he, in tears, who told us President Kennedy had been assassinated, and a few years later it was he who engaged us with reports of the atrocities of the Vietnam War.

out of the oven and sat in the chair at the horseshoe's opening. However, as this was not a usual evening, I skipped the hug and dinner plate and went directly to the chair. I didn't realize I was facing a jury; a jury composed not of representatives of my peers but of representatives of previous generations.

My Uncle was my Father's older brother, the titular head of our immigrant family; but my Father's three more years of high school caused my Uncle to defer to him. A too-long silence preceded my Uncle saying, "Jeff, I hear you're filling out papers for university."

I nodded, "Yes, Uncle Al."

My Uncle's kind eyes grew concerned as he continued, "I hear you're planning to take some courses, aah...," he looked at my Father's newspaper, "aah...phphphysics...aah... no philosophy."

"Both, Uncle Al."

"I hear you're then planning to go to law school."

"That has always been the plan, Uncle Al."

The plan for me to be a human rights lawyer was well known to my Uncle, as this plan had been consistently mentioned at family gatherings since I was an adolescent, whenever questioned about my future. Since my Grandmother's death, I took family-gathering opportunities to proselytize her human rights teachings and proudly reiterate her plan for me.

My Grandmother had been sold into indentured servitude in the Ukraine to pay off her father's debts. She was eventually sent to Siberia to support a young man through his "tenner" in one of Stalin's Gulag camps. This young man became my Grandfather. After they immigrated to Canada, fighting the injustices inflicted on Toronto's impoverished became my Grandmother's life's work;[100] work cut short by hereditary early onset breast cancer.[101] However, for a reason that would become apparent at this kitchenette conference, her social-justice work had been kept hidden from me,

---

[100] See Chapter 13, The "Helix of Life" Revisited: DNA in Concrete and Not.
[101] See Chapter 5, Princess Margaret; See Chapter 10, She Lived with the Knowledge.

likely on direction from my Father. However, at my Grandmother's funeral less than a year previously, her social-justice work could no longer be concealed.[102]

The large funeral chapel on Connaught Circle[103] just south of the University was packed with mourners. About five minutes before the service started, several mourners ran to the front of the chapel and pounded their despair on my Grandmother's closed coffin. One mourner in her grief opened the coffin and shook my Grandmother, trying to wake her. My Grandfather stood with umbrage, and ran to the coffin to protect his wife. He pushed back the mourner and other mourners who were flowing to my Grandmother's coffin. My Grandfather shouted, "Sonja worked for you all her life, let her have some peace now." During the service, I learned that my Grandmother would stand on a wooden crate in Connaught Circle, ironically across from her funeral chapel, decrying Canada's lack of concern for its starving, including recent immigrants from Hitler's concentration camps.[104] As we carried my Grandmother's simple coffin through the funeral chapel's doors, I was amazed to see a sea of mourners packed tightly together as they stood in the Circle and its spoking streets in freezing rain. Most of the mourners were under umbrellas, but as we carried my Grandmother's coffin down the chapel's steps, all the umbrellas closed in synchrony. All the men took off their hats.

My Uncle of course knew of my devotion to my Grandmother, and her plan for me to be a human rights lawyer, so I was surprised by his next question: "Are you sure you want to go to law school?"

I answered emphatically, "Yes, Uncle Al, you've known that for years."

My Uncle gave me his always caring smile before his face again dropped into seriousness. "Jeff, we want you to go to medical school."

I was shocked, and responded as quickly as shock allowed, "Uncle Al, where is this coming from?"

---

[102] See Chapter 13, The "Helix of Life" Revisited: DNA in Concrete and Not.
[103] See Chapter 13, The "Helix of Life" Revisited: DNA in Concrete and Not.
[104] See Chapter 13, The "Helix of Life" Revisited: DNA in Concrete and Not.

He ignored my question with, "School is so easy for you, and you're the only one in your generation that can get into medical school."

"But ..."

He cut me off with, "I hear you can get into medical school right out of high school."

At the time of my Uncle's question or command, some students could achieve "direct entry" from high school into medical school at the University of Toronto. I knew about direct entry because my close friend Sam[105] was applying for direct entry, but there was only one way my Uncle would know about it. My Uncle was reciting a script given to him by my Father; a script that would be my life's prescription for medicine. It was a good script. It has been a good life. It has been a gift being a physician.

My Uncle, along with his mother, brother, and sister,[106] had escaped Poland just prior to Hitler's invasion. His grandfather, for whom I am named, was the town's only doctor. He had made a deal with the town's priest and mayor to forge documents declaring his daughter and her children Christians so they could emigrate to Canada, and in exchange he would remain to continue providing the town's medical needs. Becoming Christians was necessary, as Jewish immigration was forbidden by Canada's "None Is Too Many" policy.[107] His town doctor (*felcher*) role had been inherited over many generations; generations that also included many medical school–trained specialists practising in Poland's largest cities, even Professors of Medicine in Warsaw and

---

[105] Sam's death and the subsequent prolongation of my life is described in the last chapter of this book Chapter 25, The Arrogance of "But All You Need Is a Good Index Finger."

[106] See Chapter 3, I'm Sorry Ronnie.

[107] *None Is Too Many* is the title of a 1982 book by Harold Troper and Irving Abella that brought attention to the rampant anti-Semitism that prevented Jewish immigration to Canada, the United States, the United Kingdom, and other countries. This anti-Semitism resulted in the many persons of Jewish heritage who were trying to flee Hitler's Holocaust perishing in Auschwitz and other concentration camps. The "None Is Too Many" policy in Canada was insisted by a senior Canadian official (Troper & Abella, 1982, p. xix).

Krakow. The professors in Krakow were among the first to die during Hitler's invasion, having been lined up in front of the wall beside the medical school's entrance to face Nazi firing squads. My Uncle and Father revere their grandfather, and remember him putting them on the train to their freedom while he and the rest of his family remained to soon take the train to Auschwitz. My Father is still haunted by the image of his young uncle, an athlete and his hero, running after the train pleading, "Don't let me die here." He perished along with 50 other family members. There were no survivors.

Until the kitchenette conference, I had only scant knowledge of my family's previous generations, let alone our Holocaust history. Scant knowledge in my generation was not unusual, as those who escaped Hitler rarely spoke of the Holocaust.[108] It is thought their silence is due either to survivor's guilt or the embarrassment that Jews didn't resist Hitler's roundups with rifles.[109]

My Uncle's eyes upon me were constant, as I assume were my Father's through his newspaper as if he had Superman's X-ray vision. My Mother's eyes remained bowed, staring at the pale-green plastic tablecloth, now sprinkled with tear drops. I regret not getting up and hugging my Mother, but I was too blown away by what was happening. I looked at my Uncle's eyes staring at mine. I looked at where my Father's eyes probably were.

My Uncle concluded his family history lesson with, "Jeff, you're the only one in your generation who can carry on our family tradition, so just fill out the papers and go to medical school."

---

[108] Indeed, once Holocaust refugees learned to speak English, many refused to speak Yiddish, their pre-Holocaust Jewish language. However, as my Father's father, who had been brought to Canada by Eaton's department store 10 years prior to the Holocaust to design their wedding dresses, was deaf, genetically it turned out, my family had to continue to speak Yiddish for him to lip-read. I absorbed our inherited language by osmosis rather than by my Father's intention. Even though my Grandfather was already in Canada, anti-Semitism dictated his family had to become Christians to immigrate.

[109] Akhtar, 2009; Gay, 1988; Hartman, 2014; Niederland, 1981.

"Uncle Al, you know I want to be a human rights lawyer, and medical school was never mentioned before, and—"

My Uncle cut me off sharply with, "You must go to medical school or Hitler will have won."

It was time to speak directly to my Father. "Dad, I know you're in on this, do I have any choice in the matter?"

Immigrant families are tight, but as a late-sixties teenager I felt I had choices in everything. My Father never had choices, lacking a full high school education. However, he was a math genius, and a man I respected for reasons beyond this. Yet telling me what route my life had to pursue was just too much.

My Father finally put down his newspaper. He looked at me for a moment, then sighed. "Jeff, I know you have your heart set on being a human rights lawyer, but it will be easier for you to change the world with MD after your name."

I looked at my Mother, whom I loved more than anyone in the world, and gently asked, "Mom, are you okay with this?"

She looked deeply in my eyes. She of course knew her mother's mission for me was in human rights, and she was still saying *Kaddish*[110] for her. "Jeff, I think your Father's right, you will still be able to do what *Bubie*[111] wanted you to do if you're a doctor, and you'll be able to help people in other ways, maybe you'll even find a cure for breast cancer."[112]

I went up to my room, ripped up my university application form, and the next day went to our high school's Guidance Office for a new

---

[110] *Kaddish* is the prayer for the dead that Jewish people recite twice a day throughout a year of mourning.

[111] *Bubie* is the Yiddish word for grandmother.

[112] My Mother would develop breast cancer and soon die 10 years after she said this, at about the same age as when her mother died of breast cancer. I still have not reconciled with this injustice. I tried to fulfill my Mother's wish by pursuing a cancer research fellowship. Although I contributed to the prevention of cancer of the lining of the uterus, the then second most common women's cancer, I could not make a dent in the prevention or cure of breast cancer; See Chapter 10, She Lived with the Knowledge.

one. I thank my Uncle and Dad to this day for insisting I become a physician; for insisting I not let Hitler win. I thank my Mom for saying it's okay for me not to be a human rights lawyer. I thank them to this day for the many handshakes of trust I have received from the persons I have tried to assist. I thank them to this day for the opportunity to take on a generation of national and provincial governments that did not, and still do not, see social justice as a primary imperative in health promotion and care. I thank them to this day for the gift being of a physician.

References

Akhtar, S. J. (2009). *Comprehensive dictionary of psychoanalysis.* Karnac.

Gay, P. (1988). *Freud: A life for our time.* Norton.

Hartman, J. J. (2014). Anna Freud and the Holocaust: Mourning and survival guilt. *International Journal of Psychoanalysis, 95*(6), 1183–1210.

Lee, H. (1960). *To kill a mockingbird.* Lippincott.

Mulligan, R. (Director) (1962). *To kill a mockingbird* [Film]. Universal Pictures.

Niederland, W. G. (1981). The survivor syndrome: Further observations and dimensions. *Journal of the American Psychoanalytic Association, 29*(2), 413–425.

Troper, H., & Abella, E. (1982). *None is too many: Canada and the Jews of Europe, 1933–1948.* Lester & Orpen Dennys.

Chapter 3

I'm Sorry Ronnie

Of our small generation of cousins, Ronnie was the youngest, I the oldest. We grew up together in a tight family, as immigrant families tended to be, especially families with Holocaust histories.[113] My two aunts and uncles were my parents' best friends, and I looked forward to seeing them, and of course my cousins, every weekend.

In the summers we would all go together to a small motel on Lake Couchiching. We cousins splashed away in the lifeguardless lake, while our mothers, whose swimming skills were confined to the "*mahia* stroke,"[114] stood knee-deep in the water, pleading for us to come in closer to the beach. There were no fathers on the beach, as the motel was close to a golf course. The motel was also close to a scary-looking building that we drove past each evening on our way to ice cream. The thought of ice cream, driving into town, and the lick of it coming back diverted consideration of why my parents averted their eyes as we drove by the foreboding building.

---

[113] See Chapter 2, "You Must Go to Medical School or Hitler Will Have Won"; See Chapter 10, She Lived with the Knowledge.
[114] *Mahia* is a Yiddish word that could be loosely translated as "it's a pleasure."

Ronnie was a little "different" from me, in that he spoke with his eyes, smiles, arms, and other body languages rather than the language of words. Our different languages never deterred Ronnie from running with me, kicking soccer balls with me, devouring Snickers bars with me, and, what Ronnie loved best, dancing with all of us. Indeed, if a record player was in sight, Ronnie insisted a record be playing, and all of us be dancing. The record player was usually the blonde-wood butcher-block–like RCA Victor you saw on your right as you entered my Aunt's house, or Ronnie's sister's roundish-pink "compact" record player that folded up into a small suitcase; but it could also be the dark-wood "console" in our house or in my other Aunt's house. The few records our parents possessed were 33s of Broadway musicals, with many of the songs having waltz tempos that Ronnie liked to swirl around the room to, with one of us in his arms. Ronnie's sister had rock 'n' roll 45s that the rest of us of course preferred. Thankfully, Ronnie also loved to dance to the Beatles, Stones, and Four Tops. I loved the joy in Ronnie's eyes, in his laughter, in the freedom of his dancing feet, in the pureness of his being.

Although Ronnie's not using word-language like the rest of us never deterred him, it worried his parents. My Aunt's neighbours reassured her that some children, boys in particular, develop language late, frequently offering Einstein as an example. Ronnie's doctor concurred, and further reassured that there was no need to worry. However, by the time Ronnie was seven, it was becoming evident that Ronnie would be "different" long term. His doctors used words like "mentally retarded" or worse; but no matter what labels doctors conferred on Ronnie, they couldn't diminish his joyousness, or our love for him.

Ronnie started puberty about the time I started medical school. Ronnie was developing a strong athletic body; a body that would become a curse, rather than a gift, for high school sports. Ronnie was no longer seen by his neighbours as the neighbourhood pet, but rather as the neighbourhood threat, particularly by the parents of the daughters in his neighbourhood. Fear of Ronnie in the new-world ghetto in which he lived was founded in old-world fear of persons of difference; be the difference ethnic, physical, or cognitive. A cognitive difference like Ronnie's stirred old-world superstition in his neighbours that a raven feather would soon be dropped at their door, marking them for similar misfortune. Neighbours put pressure on Ronnie's parents to have Ronnie put in a "home" with others "like him," insisting Ronnie would be happier in a "home." Ronnie couldn't be happier than he was in his own home.

The medical establishment eventually also condemned Ronnie for his difference, and suggested he be "sent away" to an institution where he could be "looked after" properly. Being "sent away" was a harsh sentence, but a sentence all too common from the medical establishment when Ronnie's fate was being debated, as Canada's health and social systems did not fund the accommodation required to enable persons of difference to remain in their homes. Being warehoused in institutions was an easier, cheaper, and "final solution."[115] Ronnie's father, a gentle man devoted to his son, did his best to help my Aunt and cousins look after Ronnie, but my Uncle needed to work outside the house to put bread on the table.

At first my Aunt categorically brushed off the thought of "putting Ronnie away." However, her face gradually became engraved with concern that maybe the neighbours and doctors were correct and she had to let go of Ronnie. I overheard my Aunt's distress in whispered kitchen-table conferences with my Father and Mother every weekend. I overheard my Father ask his sister what kind of life she could expect taking care of Ronnie, and what kind of life her two other children could expect. My Father may have been my Aunt's brother, but my Mother was her closest friend, and was more concerned with how hard it would be on my Aunt if she gave up Ronnie. After these kitchen conferences, my Mother hugged my sobbing Aunt long and hard until my Father insisted, "Let's get

---

[115] The "Final Solution" was the Nazis' term for their plan to exterminate the Jews (Browning, 2004).

going." My Aunt would reluctantly let go of my Mother, and when Ronnie turned 13, she let go of Ronnie. My Aunt never smiled again.

The institution in our province for persons "like Ronnie" was originally named The Orillia Asylum for Idiots, and eventually The Huronia Regional Centre until it was finally closed in 2009.[116] Orillia is on Lake Couchiching, the lake in which Ronnie and the rest of us splashed years earlier, and Huronia was the scary-looking building we drove past on the way to and from Orillia for ice cream. Lake Couchiching was famous as the lake in which humorist Stephen Leacock's "Mariposa Belle" sank,[117] and Orillia had become more famous as the hometown of musician Gordon Lightfoot[118] and the home of the Mariposa Folk Festival.[119] Orillia would not become infamous until years later, when testimonies from Huronia's survivors, and from family members of survivors and non-survivors, shocked our newspapers.[120]

When I first visited Ronnie at Huronia, it was as if a vacuum sucked the breath from my being. Huronia was a sad, solemn, dark place. On my way to Ronnie's "room," the medical-student me observed "the physical characteristics of Down syndrome" in persons pushing brooms in the long hallways. On entering the huge ward that was Ronnie's "room," I observed "the physical characteristics of

---

[116] Ballingall, 2016; Marlin, 2010.
[117] Leacock, 1912.
[118] "How Gordon Lightfoot's hometown of Orillia, Ont., shaped his songwriting," 2018.
[119] Mariposa Folk Festival, n.d.
[120] Marilyn Dolmage, the sister of a man with Down syndrome who died in Huronia from untreated pneumonia, described how those living there "had all of their citizenship rights stripped away.... They were lined up to eat, they were lined up to shower" (Marlin, 2010); Former residents of Huronia now participate in the Huronia Speakers Bureau, a speakers' series that tries "to ensure no one forgets—and people understand—the horrors they endured there" (Ballingall, 2016); Other survivors of Huronia work in a research project lead by Kate Rossiter that includes a theatrical production recounting their experience at Huronia (Battersby, 2018).

Down syndrome" in persons tucking in blankets on the long rows of institutional beds. Ronnie could push a broom and make his own bed at home prior to his incarceration, as likely could these other persons supposedly learning activities of daily living. As there was never a plan for these persons to return to their homes, they were in reality slaving for the institution and the province of Ontario; exploited into unpaid labour like criminals in penitentiary institutions. However, the inmates incarcerated in Huronia had nothing to be penitent for, but I would because I did nothing to free Ronnie.

When Ronnie saw me, he ran toward me and embraced me in his loving smile and warm arms. Ronnie looked as joyous as always, and eagerly took me for a walk to show me "everything." Ronnie pointed to the white-uniformed staff, to the long tables in the mess hall, to the loudspeakers, to the caged windows. As Ronnie toured me through the institution, I was surprised incarceration didn't seem to diminish the joy that was Ronnie, and I tried to convince myself that this joy might persist beyond the few hours one of us was with him. To reassure myself that Ronnie was being well looked after, I kept asking him how he was doing, but Ronnie had no words to convey what he felt amidst the injustices inflicted upon him by a system that could not accommodate him in the loving arms of his family. Over the 14 years of his incarceration, Ronnie didn't need words to convey his broken arms, blackened eyes, and omnipresent bruises that spoke beyond words.

I told myself medical school made it difficult to visit Ronnie, a four-hour round-trip car drive away. I did have labs to write up on weekends, and exams every two months requiring passing to avoid summer school; but in fact, it was my commitment to sport teams that left no time to be committed to Ronnie. It was also possible that I found visiting Ronnie difficult because I was doing nothing to address the injustices inflicted upon him. I would not until after Ronnie's death.

During the summers of my first two years of medical school, I taught swimming at a children's camp just 40 minutes from Huronia, on the lake into which Lake Couchiching empties. I sometimes visited Ronnie on my days off, but only sometimes, and only for a couple of hours; sort of squeezing him in after the laundromat and before meeting up with friends at Webers for hamburgers, or Sloan's for wild blueberry pie. I was too high on teenage helium to have it punctured by the bleakness that was Huronia.

After my second summer of a paucity of visits to Ronnie, regret impelled me to volunteer for the Metropolitan Toronto Mental Retardation Society. One of its programs had me "relieving mothers" who had sons with cognitive impairment, by taking their sons "off their hands" for two or three hours on Sunday afternoons. I could not help but notice that there never seemed to be a father in the picture, or in the pictures on the tables and walls of the small apartments where we met prior to each outing.

At our first meeting, I usually spent half an hour in the apartment gaining the confidence of the young person, and equally importantly, his mother, before I suggested, "Let's go outside and play." On the way to the door, the boys' mothers often handed me a harness with a leash attached; the harness–leash combination you see on dogs. Though I understood the mothers' concern was the safety of their sons,[121] I found the concept of a harnessed human being devastating. Ronnie was never harnessed. That first time I was handed the harness, and cognizant of not seeming to be critical of the caring mother, I suggested that if she allowed me a few minutes, I would return with something just as protective as the harness, but more fun for her son. I then dashed to a variety store and bought a colourful skipping rope. When I returned to the apartment, I asked the boy if he wanted to play a new game, and then showed him how to tie the skipping rope around his waist. I then said that I also wanted to play, and showed him how to tie the skipping rope around my waist. I tightened the knots, and we were off. From then on, I always bought a skipping rope before meeting a new young friend. It is easy to be creatively

---

[121] These boys were often termed "runners" by the medical establishment.

empowering when you are with a child needing constant observation for only a few hours. For their mothers, constant observation was constant, as illustrated in their care-etched faces. But why did I not use some of my Sunday afternoons to empower Ronnie?

After my third year of medical school, I secured a summer job at the infamous Queen Street Mental Health Centre,[122] disparagingly called "999 Queen Street." This institution was older than Canada, and its hallways were longer than football fields. The wards had massive metal doors that were always locked; indeed, locked with the enormous jail keys you see in Gothic horror movies. The thin curtains on the narrow slits to the inmates' smaller-than-jail-cell rooms were always open. The slits were once covered with jail-cell bars, the fittings for which still remained. 999 Queen Street was even worse than Huronia, but not by much.

One of my medical-student roles at 999 was performing physicals on persons on the back wards who had not seen a doctor for many years, and seemed happy to see someone carrying a doctor's black-leather bag.[123] These persons engaged me in conversation; that is, as well as they could under the cloud of massive doses of antipsychotic drugs, and sometimes electroconvulsive "therapy." Some of these persons had even had lobotomies for the purpose of control.[124]

---

[122] It was originally named the Provincial Lunatic Asylum in 1850, renamed Asylum for the Insane in 1871, Hospital for the Insane in 1905, Ontario Hospital, Toronto in 1919, and Queen Street Mental Health Centre in 1996. In 1998 it became part of the Centre for Addiction and Mental Health (CAMH) (Centre for Addiction and Mental Health, 2021).

[123] My traditional black-leather doctor's bag, a version of which has been carried by real doctors for centuries, was provided to all medical students by a pharmaceutical company. However, for more than 20 years, the Canadian Medical Association has provided first-year medical students with the doctor's bag as a backpack that is a different colour each year (Canadian Medical Association, 2021); See Chapter 5, Princess Margaret.

[124] This use of electroconvulsive therapy (ECT) and lobotomy was described in Ken Kesey's novel *One Flew Over the Cuckoo's Nest* (1962), and in the film version directed by Miloš Forman (1975) and starring Jack Nicholson, both of whom won Academy Awards for their work.

Another of my roles[125] was to check in on outpatients who lived in the rooming houses on Queen Street West, before they were evicted for gentrification purposes.[126] I thought about Ronnie frequently, wondering why I was working at 999 Queen Street rather than trying to find summer employment at Huronia.

I basically stopped visiting Ronnie when I moved two hours further south to do my residency at Western University. When I became a parent and clinician, Ronnie's predicament became even less preeminent in my mind. Indeed, Ronnie's predicament was fading from my mind until a phone call from my Father told me Ronnie had drowned in Lake Couchiching.

Huronia had organized a sleigh ride over the supposedly frozen lake, and none of the staff on the large sleigh noticed that Ronnie had fallen off the back. The next day a search team discovered Ronnie's footprints engraved in the snow, pointing to Huronia. Ronnie had

---

[125] Another role was to "attend" the medical specialty clinics in the afternoons, to which "chronic" patients were referred. I was supervised only by a nurse. In the gynaecology clinic one afternoon, the nurse suggested I perform a pelvic exam and Pap smear on a 50-year-old woman because she was reported to have vaginal bleeding. The woman's chart indicated no evidence of having been seen by a physician, let alone having had a Pap smear, since she was admitted over 10 years earlier. I entered the examining room to see a much-older-looking woman in wrist restraints, murmuring under the restraints of antipsychotics. I asked permission of the woman to gently examine her "tummy." The woman did not acknowledge this request, and I'm not sure she even knew I was there. I consulted the nurse, who nodded her head, and I went ahead and gently kneaded her lower abdomen. I was shocked by the mass I palpated there. I then asked permission to gently examine her "private parts," and with no acknowledgement from the patient, but a nod from the nurse, saw that there was no opening to her vagina, but signs the region had been irradiated. The woman must have had cancer of the cervix and was "lost to follow-up" due to her incarceration in 999 Queen St. I spoke to the supposedly supervising gynaecologist, and he referred her to Princess Margaret Cancer Hospital; See Chapter 5, Princess Margaret.

[126] These evictions were similar to the evictions for the gentrifications I refer to in Chapter 13, The "Helix of Life" Revisited: DNA in Concrete and Not, and in Chapter 14, Beneath the BMW's Wheels.

spent the night trying to walk back to Huronia until his footprints ended at a hole in the ice. A dive team was called in, and Ronnie eventually found. I was a pallbearer at Ronnie's funeral, and observed drips of water coming out of the coffin in which my cousin would be incarcerated forever. The water drips were as if Ronnie was still speaking to me, speaking as always without words.

I have a recurrent dream regarding Ronnie's drowning. We cousins are standing on a concrete pier overlooking a tumultuous lake. A large wave hits the pier and sweeps Ronnie off into water. A 15-ish me looks to the shore and sees my Mother's eyes staring at mine in distress; distress for Ronnie, but also distress for me because she knows I'm about to dive into the waves to try to save Ronnie. The next image in my dream is me swimming down and finding an underwater tunnel in the base of the pier. I swim into the tunnel and find Ronnie. His always smiling eyes are glad to see me. I smile back, grasp his hand, and start swimming him out. The tunnel soon forks into two channels, and I'm not sure which channel to choose. I swim as fast as I can holding Ronnie's hand into the channel on my left. I feel Ronnie's body go limp, and I pull him harder. I run out of breath, gasp, and wake up. I have this recurrent dream to this day. This dream has made me so claustrophobic that I avoid elevators, and refuse to be buried in the sand.[127] Ronnie's drowning was not a dream for Ronnie.

In 2013, the then-premier of Ontario, Kathleen Wynne, formally apologized for the pain and suffering of Huronia's "residents," saying the province "broke faith" by neglecting and abusing them within "the very system that was meant to provide them care."[128] The injustice of Ronnie's incarceration in Huronia continues to surface through accounts of survivors, relatives, researchers, and activists.[129]

---

[127] I describe my recurrent claustrophobic dream in Chapter 17, Ruth, and previously in my play *Calcedonies* (Nisker, 2012) and in my novel, *Patiently Waiting For...* (Nisker, 2015).
[128] Rodan, 2013.
[129] Ballingall, 2016; Battersby, 2018; Marlin, 2010.

Their descriptions of Huronia's cruelties and their efforts to remember "the horrors"[130] endured there reminisce the horrors endured in the place where Ronnie's great-grandfather and "the 50" of his family members perished, Auschwitz.[131]

Ronnie's mother died of cancer 10 years before I began writing this apology to Ronnie. I spoke with my Aunt frequently during her illness, though mostly on the phone. My Aunt always spoke at length of my Mother, her "best friend," who had died 30 years previously from hereditary breast cancer,[132] and "shouldn't have died so young." However, my Aunt never spoke of Ronnie. It was too painful. I was a pallbearer at my Aunt's funeral, and after her casket was lowered into the ground, I broke down sobbing in the soil beside it. My family comforted me, seeming surprised that I felt so much for my Aunt. I did feel so much for my Aunt, but I was sobbing for Ronnie, whose name was carved on the grave beside hers. My sobs were sobs of "I'm sorry Ronnie," "I'm sorry I didn't do more for you Ronnie," "I'm sorry I didn't speak for you Ronnie."

This narrative of injustice is my *mea culpa* to Ronnie, but it will never absolve me from my lack of advocacy for him[133] using the excuse of being too busy in medicine.

---

[130] "The horror" is a line from Joseph Conrad's *Heart of Darkness* (1899). "The horror" represents the cruelty with which the colonialist King Leopold of Belgium and his troops treated the persons in the Congo during its conquest and for a century thereafter; cruelty in the name of expanding Belgium's ivory and rubber trade (Hochschild, 1998).

[131] See Chapter 2, "You Must Go to Medical School or Hitler Will Have Won."

[132] See Chapter 5, Princess Margaret; See Chapter 10, She Lived with the Knowledge.

[133] In the five years before Ronnie's drowning, I had served on national and provincial committees, including having been Chair of the Canadian Medical Association's Committee of Specialists, and perhaps had been in a position in which I could have mitigated the treatment of Ronnie at Huronia, and maybe even saved him from drowning.

References

(2018, Nov 17). *How Gordon Lightfoot's hometown of Orillia, Ont., shaped his songwriting.* Canadian Broadcasting Corporation News. https://www.cbc.ca/archives/how-gordon-lightfoot-s-hometown-of-orillia-ont-shaped-his-songwriting-1.4895881

Ballingall, A. (2016, February 10). *Former Huronia residents join speakers' series to educate others on horrors endured.* Toronto Star. https://www.thestar.com/news/gta/2016/02/10/former-huronia-residents-join-speakers-series-to-educate-others-on-horrors-endured.html

Battersby, S.-J. (2018, April 13). *Huronia survivors work through their pain in theatre production.* Toronto Star. https://www.thestar.com/news/gta/2016/05/31/huronia-survivors-work-through-their-pain-in-theatre-production.html

Browning, C. (2004). *The origins of the final solution: The evolution of Nazi Jewish policy, September 1939–March 1942.* University of Nebraska Press.

Canadian Medical Association. (2021). Backpack program. https://www.cma.ca/cma-backpack-medical-school-tradition

Centre for Addiction and Mental Health. (2021). History of Queen Street site. https://www.camh.ca/en/driving-change/building-the-mental-health-facility-of-the-future/history-of-queen-street-site

Conrad, J. (1899). *Heart of darkness.* Blackwood's Magazine.

Forman, M. (Director). (1975). *One flew over the cuckoo's nest* [Film]. United Artists.

Hochschild, A. (1998). *King Leopold's ghost: A story of greed, terror, and heroism in Colonial Africa.* Houghton Mifflin.

Kesey, K. (1962). *One flew over the cuckoo's nest.* Viking Press.

Leacock, S. (1912). *Sunshine sketches of a little town.* John Lane the Bodley Head.

Mariposa Folk Festival. (n.d.). About & history. https://mariposafolk.com/event-info/our-history/

Marlin, B. (2010, July 26). A chance for Huronia's "invisible" to be seen and heard. *Globe and Mail.* https://www.theglobeandmail.com/news/national/a-chance-for-huronias-invisible-to-be-seen-and-heard/article1387936/

Nisker, J. (2012). *From Calcedonies to Orchids: Plays promoting humanity in health policy.* Iguana Books.

Nisker, J. (2015). *Patiently waiting for...* Iguana Books.

Rodan, G. (2013, Dec 9). *Ontario Premier apologizes for institution's alleged abuse of developmentally disabled.* Globe and Mail. https://www.theglobeandmail.com/news/politics/ontario-premier-apologizes-for-alleged-abuse-of-developmentally-disabled/article15825313/

Chapter 4

Philip

Philip follows his parent's footsteps, as they join a stream, flowing beneath "The Atrium's" glass ceiling. Philip knows too well the cacophony of this overture's coffee shops and gift shops, fundraising banners and helium-hope balloons, worried parents and hurrying staff. The Atrium's helium balloons are too-soon replaced by latex balloons painted in happy colours on the walls of the corridor to the elevators. However, Philip focuses on the poured-concrete floor's next grey line, separating stippled squares that conveyor-belt backward under his parent's shoes. Their stream slows as it merges with other streams to create a lake banked by eight elevators on its shores.

Philip's Mother turns to him, "Everything will be fine this time, Philip, I just know it will."

Philip follows his parents onto a latex-balloon–assisted elevator. His Father squeezes forward to press a button from Philip's past, "Here we go, Philip."

The elevator doors open to a too-familiar floor. Philip senses even more latex balloons are painted on the walls than last time, but Philip's eyes follow an orange line painted on the light-grey linoleum; a line Philip knows will lead him to illness. Philip forces himself to nod to the applause of "Lookin' good, Philip," "Saw your name on admissions, Philip," "Great to see you again, Philip," "Welcome back, Philip," "Even more freckles than last time, Philip."

One of the male nurses calls to him, "Hey Philip, my main man, how old are you now?"

Philip replies, "Twelve."

"Twelve? You sure you're twelve? Last time you were here, I thought you were seventeen, just small for your age."

Philip has sojourned this scape three times since illness first introduced him to this get-well warren. Philip was here for six weeks that first time, followed by four-week forays of nightly drips of promise; nightly drips of infirmity. His freckles flamed in frustration at each evening's infusion of peace deprivation, premonising "the horror"[134] of the night that will follow. Philip is reluctant to rejoin this community, to reawaken "the horror" of its nights. However, Philip feels it his duty to his parents to acquiesce to their request for another four weeks of horror.

Philip knows too well the floor's nurses and personal-service workers, food deliverers and porters. He likes best the cheerer-upper candystripers, called such because they once wore cotton-candy-pink–striped uniforms. However, since Philip has been coming here, candystripers more frequently wear other coloured stripes on the smocks that cover their blue jeans and T-shirts. Philip remembers the massive crush he had on Wednesday afternoon's candystriper the first time he was here. She wore a yellow-striped smock. Philip looked

---

[134] "The horror" is from Joseph Conrad's *Heart of Darkness* (1899), a novel that precocious Philip had read. In Conrad's book, "the horror" represents the cruelty with which the soldiers of colonialist King Leopold of Belgium, and the henchmen of ivory and rubber traders like Mr. Kurtz treated the persons inhabiting the Congo during its conquest, and for a century thereafter (Hochschild, 1998). "The horror" reverberated in Francis Ford Coppola's film adaptation of *Heart of Darkness*, *Apocalypse Now* (Coppola, 1979), in which "the horror" was perpetrated by American soldiers in Cambodia, and the words "the horror" were dramatically exhaled at the film's climax by Colonel Kurtz, portrayed by Marlon Brando.

forward all week to Wednesdays, and every minute each Wednesday, until she finally came in after school and smiled at him. At least he looked forward to Wednesdays until his doctor decided to try amphotericin. Then Philip didn't want to see her anymore, because he didn't want her to see him. Amphotericin is a cancer drug, but Philip doesn't have cancer; rather an "autoimmune"[135] condition for which "amphotericin may soon be proven effective."

Philip hears a nurse say, "Just turn left and go down to the nurses' station, Philip. I'm sure you remember where it is. You remember everything, Philip."

On the way to the "nurses' station," Philip tries hard not to notice an IV nurse pushing her cart of torture down the corridor. Philip senses she is smiling at him. In this get-well warren, even the vampires are cheery; and there are many vampires here, though the vampires here are different from movie vampires in that they rise at dawn rather than disappearing into their coffins, wear white rather than black, and are more interested in the veins in your arms than the veins in your neck. It must be hard for the vampires to keep cheery, having to bite kids every morning, and because just like when movie stars see vampires, the children scream even before they're bitten. Philip understands that someone has to do these evil deeds, and is always pleasant with the vampires, even speaking vampirese with them. When a vampire says, "*Guten morguening,*" Philip responds, "*You vant to bite mine arm? Gut vit me.*" His Count von Count accent makes even the terrified kids in his room smile because they know Count von Count from *Sesame Street*. However, Philip's Count von Count gradually weakens over the weeks of amphotericin.

Philip knows that the paediatricians are not necessarily cheery here, nor their residents, nor the medical students—at least when their supervisor is around. When they are alone with Philip, some

---

[135] In autoimmune conditions, a person's own antibodies attack their cells (Brower, 2004). Autoimmune conditions in young person's include juvenile rheumatoid arthritis and juvenile dermatomyositis.

joke around a bit. Last time he was here, one of the residents called Philip "Einstein," and another called him "Motor Mouth." But mostly they just hit on the nurses.

A nurse says, "Follow me, Philip, I'll get your chart. I think you're in the same room as last time. Right, let's go, Philip."

It is the same room as last time, but not the same bed. He tries not to let his shoulders melt in the fluorescent shadow of the IV tree growing from the head of what will be his bed; a tree that will too soon dangle unforgiving fruit.[136] Philip sits on the edge of the bed. His parents stand staring at him for a while, and then surreptitiously staring at each other. Philip does not challenge his Mother's parting prognosis, "Everything will be fine this time, Philip, I just know it will." Or his Father's, "Remember how confident your doctor is in the new side-effect drug." Philip knows that their reassurances are as much for them as they are for him.

As soon as his parents leave, Philip slides off the edge of the bed, goes to its foot, and cranks the bed's head up. Philip takes *The Odyssey* out of his backpack and places it on the bedside table. He then sits on the bed, kicks his shoes off, and settles in to read just as a nurse comes in. "Need to take your temperature, Philip."

Philip sucks on the thermometer like a lollipop.

"Very funny, Philip, you know the drill."

Philip wobbles it around in his mouth, and then lets it hang out like Joe Camel[137] smoking a cigarette.

"C'mon, Philip, keep it still."

---

[136] I am reminded of Billie Holiday's (1939) powerful song, "Strange Fruit," when I see chemotherapy bags hanging from IV trees. Billie Holiday witnessed Black men and women hanging from trees in the "Jim Crow" American South, "lynched" by the Ku Klux Klan and men who did not believe they needed the camouflage of white hoods to commit murder. "Strange Fruit" comes to mind even though the chemotherapy bags are hung out of hope, while the Black men and women were hung out of racial hatred.

[137] Joe Camel was a cartoon-like friendly camel on each package of Camel cigarettes. Joe Camel's features were specifically designed to addict boys to smoking (Pierce, Choi, Gilpin, Farkas, & Berry, 1998).

Last time Philip was here puking his guts out, one of the nurses said he was an "intrepid young man." Philip actually had to look up "intrepid" in his Pocket Dictionary. Good word. Maybe his wanting to remain "intrepid" is why he's back. That and his parent's encouragement. There's a beep. Philip takes the thermometer out of his mouth, is about to hand it to the nurse, but takes it back for one last lick.

"Temperature's normal, Philip. Have a good night."

The ward quickly softens to sleep. The kids in the three other beds nestle into their pillows. Philip assumes they are much sicker than he is. They are definitely much younger. Their beds have high railings on their sides, caging them in. They also have musical mobiles dangling above their heads; carousels of jungle animals, farm animals, birds, fish. The hospital silences the mobiles' musical mechanisms to avoid the discordance of conflicting tinklings. Conflicting tinklings would be better than silence of anticipation.

An IV nurse comes in. "Nice to see you again Philip," she whispers as she hangs a clear "normal saline" bag and its loops of IV tube from the IV tree. Philip thinks, I wish I could say the same for you, but whispers, "Nice to see you again too."

"Need to start your IV, Philip, so you can start your medicine." The nurse puts on sterile gloves, then places a sterile towel on Philip's lap. She deftly unwraps an IV-insertion needle, complete with white plastic "butterfly," then alcohol-swabs the back of Philip's left hand where a "good vein" might be located. A "good vein" becomes harder to find over time, as amphotericin is toxic to veins, just as amphotericin is toxic to the rest of a person. The IV nurse taps the back of Philip's hand to "stand up" his veins. Nurses prefer inserting IVs into back-of-hand veins to permit greater mobility for the patient, even though hand veins are small and difficult to "hit." Nurses prefer inserting IVs into the opposite hand from the hand with which a patient writes, or colours with for that matter. Philip silently pleads that a vein will stand up; however, his left hand's veins remain hidden. The nurse is having trouble. "I'll be right back, Philip."

A few minutes later the nurse returns with a warm towel. She tests its temperature so Philip's skin won't be burned. Philip doesn't care if his skin is burned by her towel, as long as the IV insertion is perfect, so the amphotericin won't go interstitial again and burn his skin from inside out.

The nurse decides to try the back of Philip's right hand. She places the warm towel on it for a minute, alcohol-swabs, and taps. The nurse soon gives up on Philip's hand veins, and starts warming, alcohol-swabbing, and tapping just above Philip's left wrist. A vein there promises to co-operate. The nurse stares at her chosen target, alcohol-swabs it, and punctures; but she's not in. She jiggles the needle. The IV insertion hurts just as much as last time. She smiles sympathetically. She's having trouble, just like another IV nurse did last time.

"Sorry, Philip," she whispers.

The nurse tries another vein. Philip tries to suppress sounds of pain. The nurse knows that Philip knows she's having trouble.

"Almost got it, Philip," she whispers. "Sorry. I'll try your right wrist."

The nurse warms, alcohol-swabs, taps, and says, "I see a good one, Philip." She punctures and smiles, "Got it, Philip, don't move." The nurse quickly flips off the plastic sheath from the IV tubing's needle and plugs it into the rubber nipple of the white plastic "butterfly" insertion-needle, then quickly adhesive-tapes the "butterfly" to Philip's wrist. She turns a tiny white plastic cog wheel near the top of the IV tube that permits the drips of clear fluid. She tells Philip to keep his wrist straight for a minute because "the IV's temperamental." The nurse tapes a plastic splint to the front of Philip's wrist. "That should do it, Philip," she whispers. "Have a good night."

Philip watches the salty fluid drip from its IV bag into the clear plastic reservoir at the top of the IV tubing. The drips premonisce what will come. Too soon, another nurse comes in. This nurse will "hang" the amphotericin. She smiles encouragingly.

"How's it going, Philip?" she whispers.

"Fine, thank you," Philip whispers back.

The nurse first hangs another clear-fluid plastic bag, this one containing "electrolytes" to replace the electrolytes that Philip will barf into the aquamarine "k-basin" sitting ominously beside *The Odyssey* (Homer, 1614/1967) on his bedside table. The "k" in k-basin stands for kidney because the basin is kidney-shaped so that its curve can fit under your neck when you barf. The aquamarine colour is an offensive reminder that the barf-basin resembles a kidney-shaped swimming pool. The nurse plugs the needle from the electrolyte bag into the nipple she has just alcohol-swabbed on Philip's IV tubing.

The nurse's encouraging smile continues, even as she hangs the Dijon mustard–coloured amphotericin from the IV tree. The nurse then alcohol-swabs another of the rubber nipples on Philip's IV's tubing, and spears in the amphotericin.

"Your IV's running well, Philip," she whispers.

The nurse's thumb then rolls open amphotericin's flow. Poisonous mustardy drips stain the clear fluid in the plastic chamber beneath its bag. The nurse's encouraging smile continues as she hangs a white bag that Philip has not observed previously. The white in the bag must be the experimental anti-nauseant drug that they hope will slay amphotericin's sins. The nurse swabs the final nipple, plugs in hope, then quickly adjusts the "drip rates" from the bags.

"Here we go, Philip," she whispers. "See you tomorrow night."

She turns to go but stops, wiggles the big toe of Philip's right foot, and says, "Everything will be fine, Philip."

Philip tries not to stare at the mustardy drips as they slowly tick from the amphotericin bag to the explosion that will soon come. He stares at the ceiling instead, where thousands of tiny black holes stare back at him from the greying rectangular panels that hide wires, cables, and pipes. A panel was removed from its metal frame last time Philip was here, and a man wearing orange coveralls along with a spaceman-like white hood and clear plastic visor climbed up a ladder to shine his large square flashlight into the snake-filled abyss. A nurse had herded Philip and the others out of the room, but Philip watched from the doorway. Philip now stares harder at the holes in the ceiling panels. Their patterns seem completely random. Philip also notices tiny faint lines in the panels that he had never noticed before. The lines remind Philip of contour lines on a topographical map.

Philip still hears the words of the nurse who hung the hoped-for side-effect soother echoing off the walls: "Everything will fine, Philip." Philip hopes that the nurse might have seen the new anti-side-effect drug work successfully in another patient. She might even know for certain that the new anti–side-effect drug will make this "course" of amphotericin "tolerable." "Course" is doctor talk for Philip receiving amphotericin every night for three weeks. "Tolerable" is doctor talk for not puking your guts out every night.

Philip hears another nurse coming in. "Just making rounds," she whispers. "Is everything okay, Philip?"

"Yes, thank you," he whispers back.

Philip again notices the no-longer-musical mobiles above the beds of his roommates. Philip wishes he was still young enough to have a mobile cheering the air above his bed, even if its music was turned off. Spaceships would be best, so he could dream of leaving. But instead of mobiles, Philip tilts his head up to witness four IV bags; two clear, one white, one a menacing murky-mustard. All bags void of melody and mirth; one filled with fear and foreboding.

Midnight passes on Philip's Timex Ironman that his Father gave him when he started high school last fall. Philip is in a special program to make things "more challenging." At least the novels are better.

At 2:15 a.m., the first wretched ripples reach Philip's shore. Soon enormous waves of ache crash to his core, cannonballing his supper into the kidney-shaped pool he grabbed just in time. Philip tries to contain the splashing within its aquamarine rim, but brownish green splatters the white bedsheets. Between heaves, he tries to mop up his vomit with a facecloth.

Philip wonders if he should press the button for a nurse. A nurse's hands could dim amphotericin's sins for a while. No, he's too old for that, and there are only two nurses on the night shift because of the "funding cutbacks" he overheard the nurses talking about.

Only the white adhesive tape of his wrist-shackle keeps Philip here. That and his promise to his parents to try again. He barfs again. His k-basin is overflowing. He presses the button for a nurse.

A nurse soon enters the room. "Sorry, Philip," she whispers. "Let me clean things up a bit. You're doing fine, Philip."

She gives Philip another k-basin.

"I'll come back soon," she says.

The raging storm drags Philip to dawn's calm. He hears the cheery chirpings of the next shift at the nurses' station. Soon after, Philip hears nurses' "good mornings" in rooms down the hall. There's nothing good about Philip's morning. He's exhausted. Nothing will be good about Philip's day. He'll even be too tired to read *The Odyssey*.[138] He was also hoping to read *Great Expectations*[139] and *The Grapes of Wrath*[140] during this internment; however, his hoping was contingent on the anti–side-effect drug working. Now the presence of *The Odyssey*[141] on his bedside table, along with the other books in his backpack, just adds extra heaviness to his day.

Philip tries to lift his head to greet the woman delivering breakfast, but he can't. She looks sympathetically at Philip's eyes, as she places his tray on the over-the-bed table she has slid into position. Philip works up a smile, and is about to say, "*Nada*," when he sees a vampire hovering over her left shoulder, eager to take his blood for an electrolyte check. Philip closes his eyes in surrender.

Philip hears, "*It von't hurt a bite, Philip.*"

The IV nurse is trying to get a smile by imitating Philip's vampire accent, but she doesn't quite get the Count von Count correct. She needs to watch *Sesame Street* with the kids on the other side of his room for a few mornings. Philip dutifully puts pressure on her puncture hole for the prescribed almost two minutes, then decides to get out of bed to try to start his day.

---

[138] Homer, 1614/1967
[139] Dickens, 1861/1975
[140] Steinbeck, 1939/2014
[141] Homer, 1614/1967

As Philip is brushing his teeth, he sees his doctor enter the room. His flowing white coat is followed by a clothesline of white coats, worn by the younger doctors and medical students. They circle like swans around the bed across the room nearest the door, stay for a few minutes, then circle the bed beside it. Philip's doctor always goes to the foot of the bed, while the "fellow" presenting the "case" goes to its head, and the residents and cygnets quickly circle in. The word "fellow" is a misnomer, as the "fellows" are usually women, but this one happens to be a man. Knowing the drill, Philip returns to his bed before "the team" flocks around it, not wanting to slow it down. "The team" soon floats over.

The fellow starts presenting Philip's "case": "This 12-year-old boy was admitted yesterday for amphotericin. His pre-admission blood work indicated that we could start treatment. He—"

Philip's doctor interrupts. "Philip, I heard you were admitted yesterday. How have you been since I last saw you last?"

"Fine, Doc."

The fellow continues, "I wrote his medication orders yesterday afternoon so he should have had his first infusion of amphotericin last night. Yes, I see the empty bag."

Everyone stares over Philip's head at the mustard-tinged bag hanging from the IV tree; that is, except Philip's doctor and Philip. Philip actually makes a point of not looking at the wretched bag until a nurse comes in and disposes of it; into the hospital's incinerator he hopes, or at least to a dump with the hospital's radioactive materials and toxic-chemical waste. Philip's doctor looks at Philip sympathetically.

"How did you manage last night, Philip?"

"I can handle it, Doc."

The doctor turns to his learners. "In addition to amphotericin, Philip is receiving an experimental drug to ameliorate the side effects of amphotericin. It has had very promising results in clinical trials with cancer patients."

They all nod, pretending interest.

His doctor continues, "Philip doesn't have cancer, and we were fortunate to get this drug off-trial for him."

They all nod again. A cygnet asks a biochemistry question to impress Philip's doctor. It works. His doctor beams, and fires off a mini-lecture in biochemistry-language on the theory of how the anti–side-effect drug should work. However, unless Philip's barfing woke up the other kids, he's the only one in the room who knows the anti-barf drug doesn't work.

"Any other questions?"

One of the medical students smiles warmly at Philip, noticing *The Odyssey* (Homer, 1614/1967) on Philip's bedside table. He asks Philip to tell him about the books he likes to read. The other members of "the team" seem disgruntled by this medical student's delay, except for his doctor, who smiles and nods at the medical student. Philip quickly tells the student about his favourite books: *To Kill a Mockingbird*, *The Hunchback of Notre-Dame*, *A Tale of Two Cities*, and the book he's currently reading, *The Odyssey*.

Philip's doctor is about to turn and leave, but looks at Philip's eyes and again asks about last night. Philip chins, "I can handle it, Doc." His doctor seems unsure, and sort of smiles at Philip as he hesitantly moves on to the next bed. The cygnet circle quickly dissolves to recrystallize around the bed beside Philip's. After "the team" leaves his room, Philip silently encourages himself with, "I can handle it, I can handle it;" but hears his lips quiver, "Not another amphotericin night."

Philip is exhausted. His head collapses back on the pillow. In an hour or so, Philip hears the breakfast-server coming back. He hears her removing his tray, then putting it back on his over-the-bed table. "You must eat, Felipe."

She knows his name from his card on her food-cart and the matching name-card at the foot of his bed.

"You must eat to get well, Felipe."

Philip lifts his head long enough to say in *Sesame Street* Spanish, "*Gracias, no tengo hambre, Señora.*"

Excitedly she asks, "*Hablas Español, Felipe?*"

"*Nada.*"

"*Muy bien. Veo, tu peudes hablas un poquito de Español. Por favor, come algo tienes que mantenerte fuerte.*"

Philip opens his eyes. She has walked to the head of his bed, and is staring into his eyes.

"*Por favor*, try to eat Felipe." She smiles a concerned smile, and leaves.

Philip tries to go back to sleep but can't. Hospital noises are amplified when you're tired. He gets out of bed and aimlessly wanders the halls. The kids in wheelchairs lining the hall are sicker than he is. He tries not to meet the eyes of any staff because he doesn't have the energy to smile, let alone have a conversation. A nurse comes right up to him. He avoids her eyes, but she touches his shoulder and asks, "Do you want to go to the craft room, Philip?"

Philip smiles, "No, thank you."

"What about the book cart?"

Philip smiles again, "No, thank you. I've brought several books from home."

The nurse returns Philip's smile and proceeds down the hall. Philip returns to his room to try to read, but can't. He tries to draw, but has similarly frustrating results. He doesn't even attempt to write. Too soon it's "Visitors' Hours."

Philip tries to hide last night's horror from his Mother when she comes in looking worried and asks, "How did it go, Philip?"

Philip tries to hide last night's horror from his Father when he hesitantly asks, "Did the new anti-nauseant drug work, Philip?"

Philip knows they both know it didn't work because they keep staring at the dark circles masking his eyes and surreptitiously glancing at each other's eyes. Of course, his parents can't admit they know, for that would admit they know what he will go through tonight.

Philip doesn't have much to say, which is also worrisome for his parents. He tries to make conversation, knowing they can't stay long anyway, as it's more than a three-hour drive home. Finally, Philip hears "Visitors' Hours are over." His Father says, "We have to go, son," squeezing his Mother's arm. His Mother says, "We love you, Philip, we'll see you tomorrow." Philip exhales a smile of relief.

The grey brick outside the window is starting to redden. Philip hopes the reddening will last forever, holding off the amphotericin. The friendly medical student comes in.

"Philip, I'm rotating off Paediatrics tomorrow, but I was hoping I could keep visiting you to talk about books." He looks at Philip's eyes. "If you're feeling up to it of course." Philip nods assent. The medical student tells him, "I rent a room in the St. James Town apartments near Cabbagetown,[142] an easy jog to the hospital."

Philip thinks that the medical student is trying to make him laugh, as his sister has a Cabbage Patch doll.

The medical student points to *The Odyssey* languishing on Philip's bedside table. "I just happened to buy that book today; coincidence of course." He smiles at Philip, and Philip smiles back.

"I was thinking of going back to the bookstore for more fiction Philip, any suggestions?"

Philip has depleted his small reserve of smiles and just shakes his head. The medical student understands. "Philip, I hope you have as good a night as possible." The medical student is soon replaced by the nurse who will hang the amphotericin bag and the lack-of-vomit-protection bag. Philip tries to hide his fear from her, as well as from himself. Too soon he has to surrender to amphotericin's aggression arching his being.

The next morning, when the swan flotilla flocks his bed, he retorts his doctor's encouraging enquiry of "How was your night, Philip?" with, "Actually, not so hot, Doc."

Philip had been determined to hide "the horror" from his doctor, but finds himself asking, "Doc, isn't there an alternative to amphotericin? I mean it's not like I have cancer."

His doctor pauses, begins to speak, then pauses again. He puts his hand on Philip's right foot, "No, son, you don't have cancer."

---

[142] See Chapter 13, The "Helix of Life" Revisited: DNA in Concrete and Not; See Chapter 14, Beneath the BMW's Wheels.

"Then let me try something else."

"I'm sorry, Philip. I wish there was something else. As it is, amphotericin is experimental for your condition."

"I read that there could be genetic therapy for some people. Let me try that, even if it's also experimental."

"I'm afraid not, Philip. You can handle this."

Philip knows he cannot.

"Then maybe less amphotericin."

"I'm sorry, Philip, it won't work then."

"That's okay, Doc, just thought I'd ask."

"Sorry, Philip. See you tomorrow."

This day seems longer, probably because Philip is too tired to push his IV tree around the halls; too tired to even read. His parents enter his room, after stopping to press their smile buttons just outside the door.

"Hello my young man," says his Mother.

"I see you're keeping your chin up, Philip," says his Father.

Philip feels like saying, "That's because I pressed my smile button too," but of course he doesn't, there's no point.

"I'm hanging in."

This will be the last time his parents will visit him until the weekend. They can't keep come down every evening. Philip tries to keep smiling but can't. So, he tells them, "I want to try to sleep for a bit."

"We'll watch you then," his Mother says.

"I appreciate you coming down to see me, but you should get an early start home."

"We'll get a coffee and come back later," his Father says.

Too soon, his parents return. Too soon, it's "Visitors' Hours are over." Too soon, the amphotericin nurse comes in. She's thinking about changing Philip's IV site. She's worried that if the amphotericin leaks "interstitially" it will burn his skin. The nurse decides the IV will last another day.

"I know this is tough on you, Philip," she says. "Is there anything I can do?"

Philip thinks, "Yes, three things. First, don't plug your poison into me; second, say goodbye; third, never come back." But instead he says, "I'm afraid not." She plugs in the amphotericin, and says, "Good night."

This night seems longer. Philip tells himself he must accept an amphotericin-alloyed future, but as the amphotericin bag drips its venom through his being, and the anti–side-effect bag drips its incompetence beside it, and as runnels of saltwater drip into his mouth, Philip changes his script. He doesn't have to accept this.

Suddenly Philip feels much better, as if the experimental anti–side-effect drug is actually working. Of course it's not working, but Philip giddily accepts the amphotericin, and the wretchedness it is wrecking upon him. Philip begins taunting the amphotericin bag, "Hurt me tonight if you want to because it's your last chance. Tomorrow, when I see my doctor, I will tell him I refuse you. Tomorrow night I will sleep. The next day I will read again, I will draw pictures again, I will smile again, I will be Philip again."

Philip welcomes the dawn. He has spent the last hours of darkness rehearsing his refusal of amphotericin. His dress rehearsal occurs as the white coats tighten around the beds across the room. When they come over and circle his bed, his doctor asks, "How did it go last night, Philip?"

Philip inhales, slowly nods twice, exhales, "Doc, I had my last ampho-terrible night."

A short preliminary version of "Philip" was published in the *Canadian Medical Association Journal*,[143] and a short theatre version was published in *From Calcedonies to Orchids: Plays Promoting Humanity in Health Policy*.[144]

---

[143] Nisker, 2003.
[144] Nisker, 2012.

References

Brower, V. (2004). When the immune system goes on the attack. *EMBO Reports, 5*(8), 757–760.

Conrad, J. (1899). *Heart of darkness.* Blackwood's Magazine.

Coppola, F. F. (Director) (1979). *Apocalypse now.* United Artists.

Dickens, C. (1975). *A tale of two cities.* Penguin Books. (Original work published 1859)

Dickens, C. (1975). *Great expectations.* Heinemann Educational Books. (Original work published 1861)

Hochschild, A. (1998). *King Leopold's ghost: A story of greed, terror, and heroism in Colonial Africa.* Houghton Mifflin.

Holiday, B. (1939). Strange fruit [Song]. Commodore.

Homer. (1967). *The odyssey* (A. S. Cook, Trans.). Norton. (Original work published 1614)

Hugo, V. (1987). *The hunchback of Notre-Dame.* Penguin Group. (Original work published 1831)

Lee, H. (1960). *To kill a mockingbird.* Lippincott.

Nisker, J. (2003). Philip. *Canadian Medical Association Journal, 168*(6), 746–747.

Nisker, J. (2012). *From Calcedonies to Orchids: Plays promoting humanity in health policy.* Iguana Books.

Pierce, J. P., Choi, W. S., Gilpin, E. A., Farkas, A. J., & Berry, C. C. (1998). Tobacco industry promotion of cigarettes and adolescent smoking. *Journal of the American Medical Association, 279*(7), 511–515.

Steinbeck, J. (2014). *Grapes of wrath.* Viking. (Original work published 1939)

Chapter 5

Princess Margaret

The injustice of children being ill made all eight weeks on Paediatrics was tough, but my two-week rotation at Princess Margaret Hospital[145] was just too much. As cancer is predominantly an adult disease, the old Princess Margaret was an adult hospital, lacking the brightly painted walls, colourful helium balloons, and animated wall-decals of the children's hospital where I had worked the previous six weeks.[146] Children with cancer had to go to Princess Margaret because it was the location of the massive radiation machines.

Three years prior to my rotation at Princess Margaret, I had the privilege of learning radiation physics from Dr. Harold Johns, the physicist who developed focused radiation.[147] Radiation physics so

---

[145] In 2012, Princess Margaret Hospital was renamed Princess Margaret Cancer Centre (University Health Network, 2021).

[146] My previous "rotation" had been at Toronto's Hospital for Sick Children; See Chapter 4, Philip.

[147] Dr. Johns and his clinical counterpart, the radiotherapist Dr. Ivan Smith, had worked at the University of Saskatchewan when they made this development. Dr. Johns later moved to the University of Toronto, where he taught me, and Dr. Smith moved to Western University, where I spent considerable time in his Unit during my residency, and then later as a clinician and cancer researcher.

enamoured me that I could not stop reading Dr. Johns's thick textbook.[148] In the 2018 biopic *First Man*,[149] a referral to Dr. Johns was requested by astronaut Neil Armstrong for his three-year-old daughter dying of cancer. Although I was enamoured by Dr. Johns and his textbook, I was not enamoured by the thought of having to go to Princess Margaret, where my Grandmother[150] had been subjected to Dr. Johns's radiation machines for metastatic breast cancer, and where, in the last months of her short life, she endured repeated abdominal "taps" to remove the cancerous ascitic fluid that expanded her abdomen inhibiting her breathing.[151] My Mother would also have radiation for her early onset breast cancer at Princess Margaret, but it would be the new Princess Margaret across the road from the Medical School's new building.[152]

During my Grandmother's months of daily trips to Princess Margaret, she lived in a hospital bed in my room,[153] while I bunked next door with my Brother. My Grandmother felt terrible about displacing me, and worse about the need for my Father to have to drive her to Princess Margaret every morning and pick her up there every evening. In the funeral limo going to the cemetery to bury my Grandmother,[154] my Father told me my Grandmother begged him to throw her off the Bloor Street Viaduct[155] as they drove across it each

---

[148] Johns, 1983.

[149] The film's title, *First Man* (Chazelle, 2018) refers to first man to step on the moon, Neil Armstrong, who was portrayed by Ryan Gosling.

[150] See Chapter 10, She Lived with the Knowledge; See Chapter 11, Dr. King, The Little Prince, and Seeing with One's Heart in Medicine; See Chapter 13, The "Helix of Life" Revisited: DNA in Concrete and Not; de Saint-Exupéry, 1943.

[151] See Chapter 10, She Lived with the Knowledge: See Chapter 13, The "Helix of Life" Revisited: DNA in Concrete and Not; Nisker, 2004, 2012.

[152] See Chapter 13, The "Helix of Life" Revisited: DNA in Concrete and Not.

[153] See Chapter 10, She Lived with the Knowledge; Nisker, 2012.

[154] See Chapter 10, She Lived with the Knowledge; See Chapter 13, The "Helix of Life" Revisited: DNA in Concrete and Not; Nisker, 2012.

[155] There were no safety screens then on the Bloor Street Viaduct making it a common site for suicide, including that of one of my teenage friends. The

day on the way to and from Princess Margaret. My Grandmother preferred to die rather than inconvenience anyone. Eight years later, as I drove across the Bloor Street Viaduct to re-enter Princess Margaret's doors, I knew Princess Margaret would reminisce the injustice of my Grandmother's "too-young" death,[156] but I had underestimated the injustices that awaited me there.

I was assigned a ward of boys, six to ten years old, whose 16 beds formed a square perimetering the room. None of the children had hair on their heads. Instead of hair, their scalps bore dark-blue plus signs, tattooed as a target for radiation to their brains. All the children had vacant expressions in their eyes, expressing the confusion of what was happening. The injustice of their suffering was beyond my capacity to "limit emotion," though I had been instructed to "limit emotion" and never let my heart get in the way of "objective" medicine.[157] These children could never be objects with whom I could be objective; they would always be innocent young persons caught in the injustice of cancer's wrath. My heart was definitely going to get in the way of being objective with these children, and I was going to let it.

My prescribed function on the Ward was to perform "histories and physicals" on the children; "histories and physicals" that were already documented on their charts in compulsory blue ink by the last medical student or two at Princess Margaret. There was no way that I would prod any of these young persons into giving me their "history of present illness," let alone prod their abdomens searching for an

---

bridge was originally erected as a "viaduct" to transport water pipes across Toronto. Michael Ondaatje describes the building of the bridge in *In the Skin of a Lion* (Ondaatje, 1987). The Bloor Street Viaduct is also featured in the Barenaked Ladies' song "War on Drugs" (Barenaked Ladies, 2003) and Bruce Cockburn's song "Anything Could Happen" (Cockburn, 1988).

[156] See Chapter 10, She Lived with the Knowledge.
[157] See Chapter 3, I'm Sorry Ronnie; See Chapter 7, I'm Sorry Vaccine Came Too Late for You Janet; See Chapter 11, Dr. King, The Little Prince, and Seeing with One's Heart in Medicine.

enlarged liver or spleen that a previous prodder had already detected, and too-often had delineated in blue ink on the child's skin. It made no sense to prod them again; to further hurt young persons enduring cancer, radiation, chemotherapy. It made no sense to hurt them again for the sole purpose of having a tick on my graduation report card indicating competence[158] in taking histories and performing physicals on paediatric oncology patients.

On my first day at Princess Margaret I did prod the children, but only to ask their names as I sat at the non-railinged foot of their beds in my freshly laundered white coat. My stethoscope remained in the black-leather doctor's bag provided to all medical students by a pharmaceutical company.[159] On my second day at Princess Margaret, instead of carrying the black-leather doctor's bag, I carried a black-cardboard guitar case, inside of which was the acoustic guitar I had purchased at a pawn shop just south of Princess Margaret. After waving enthusiastically at the children, I lifted a chair into the middle of the Ward and placed my carefully washed guitar case upon it. Then, with an *abracadabra*, I opened my guitar case and took out my carefully washed guitar to smiles, trying-to smiles, and just-too-weak-to-try-to smiles. However, all eyes were opening a bit wider. Some of the children sat up. Some of the children were too ill to sit up.

After wrapping my red-and-black mock-mosaic guitar strap around my neck, I walked around the room coaxing the children to tell me what songs they wanted to sing. I fully intended to spend the day singing with these children, all of whom I was sure would benefit more from song than from a medical student's stethoscope on their

---

[158] Competency-based medical education (CBME) (Royal College of Physicians and Surgeons of Canada, 2021).

[159] My traditional black-leather doctor bag, a version of which had been carried by real doctors for centuries, was provided to all medical students by a pharmaceutical company. However, for more than 20 years, the Canadian Medical Association has provided first-year medical students with the doctor's bag as a backpack that is a different colour each year (Canadian Medical Association, 2021).

chest, or a medical student's hard-rubber reflex hammer hammering their knee, or a medical student's fingers digging for their liver and spleen. I was committed to using the music of my guitar, along with happy-children song lyrics, to distract sickness with song and lift spirits with the helium of singing.

The songs we sang always had choruses after each verse that were easy for the children to learn quickly. I sang the same songs every day, so many of the children were soon able to sing the verses as well as the choruses, and without my voice's assistance. In a few days, I heard excited requests for particular songs, and a touch of joyousness in the song requests. Over the two weeks of my Paediatric Oncology rotation, I confronted the imperative to "limit emotion" with song, knowing that song might be doing at least some good for these children, who I knew were doing more than some good for me. I was learning much from these children without ever practising on them; learning courage and kindness, generosity and acceptance, all of which would assist me in soon being a good physician, and years later a good patient.[160]

Even though these children were teaching me to be a good physician, and becoming a good physician is what medical school is supposed to be about, my supervisors at Princess Margaret were not impressed with my performances, nor with my performance on my Paediatric Oncology rotation. I was failing to deliver to the charts of my pretend-patients the "histories and physical findings" and the "investigations and treatment plans" that my supervisors wanted to see documented in blue ink above my blue-ink signature; even though they were already documented above the blue-ink signatures of other medical students. One supervisor insisted, "Buck up and do what clinical clerks are supposed to do." Another supervisor, more sympathetic to my thinking-with-my-heart,[161] pleaded, "At least do a

---

[160] See Chapter 25, The Arrogance of "But All You Need Is a Good Index Finger."
[161] See Chapter 11, Dr. King, The Little Prince, and Seeing with One's Heart in Medicine; de Saint-Exupéry, 1943.

few histories and physicals so that you'll pass your rotation." I of course could not.

I learned much at Princess Margaret, probably more than on any other rotation in my clinical clerkship, but received a failing grade for my two weeks on Paediatric Oncology. This grade was meaningless to me as my overall grade in my eight weeks on Paediatrics would be good enough for me to pass, and much more importantly, I could not assist these young persons in any way but through song.

References

Barenaked Ladies. (2003). War on drugs. On *Everything to everyone*. Reprise.

Canadian Medical Association. (2021). Backpack program. https://www.cma.ca/cma-backpack-medical-school-tradition

Chazelle, D. (Director) (2018). *First man*. Universal Pictures.

Cockburn, B. (1988). Anything can happen. On *Big circumstance*. True North Records.

de Saint-Exupéry, A. (1943). *The little prince*. Harcourt Brace Jovanovich.

Johns, H. E. (1983). *Physics of radiology*. Charles C Thomas Publisher, Ltd.

Nisker, J. (2004). She lived with the knowledge. *Ars Medica*, 1(1), 75–80.

Nisker, J. (2012). *From Calcedonies to Orchids: Plays promoting humanity in health policy*. Iguana Books.

Ondaatje, M. (1987). *In the skin of a lion*. McClelland and Stewart.

Royal College of Physicians and Surgeons of Canada. (2021). What is competence by design? https://www.royalcollege.ca/rcsite/cbd/what-is-cbd-e

University Health Network. (2021). Princess Margaret history. https://www.uhn.ca/corporate/AboutUHN/OurHistory/Pages/princess_margaret_history.aspx

Chapter 6

Miriam

The evening I met Miriam, she was sitting on the edge of a hospital bed in a two-bed room, wearing a standard light-blue hospital gown. Miriam forced a smile to return my smile, looked at my name tag, and said, "You're going to think I'm crazy, but I know I'm going to die soon if you don't take out my ovaries." I assured Miriam, "You're not going to die of ovarian cancer," before pulling the no-longer-white hospital curtains around her bed. Then from a bedside chair I asked Miriam her to please tell me her family's history of cancer. She looked deep within my eyes, exhaled, nodded her head, and handed me a wrinkled sheet of yellowing paper with her family tree drawn on it in smudged-lead pencil.[162]

Earlier in the day, I had been beeped to the office of my mentor, a famous cancer surgeon and the reason I chose Western University for my residency. For some reason he looked uncomfortable, not leaning back in his large tan-leather chair with his hands behind his head smiling at me as he usually did when I visited his office. Instead he stood with seriousness as I entered and put his huge hands on my

---

[162] I prefer the term "family tree" to the medical (and for that matter veterinary) term "pedigree" when referring to a diagram illustrating a family history.

shoulders. "Jeffer, I just admitted a 39-year-old woman to our service to remove her ovaries, purely on her request, because she has a morbid fear of ovarian cancer due to her family history, even though I assured her ovarian cancer is not hereditary." He continued in an apologetic voice I had not previously heard, "Jeffer, unnecessary surgery is inexcusable, but this woman says she can't cope because she is sure she will die soon of ovarian cancer, and leave her young children motherless." He took a deep, almost frustrated breath. "Jeffer, please see her and tell me what you think."

Miriam had been referred to my mentor by the third surgeon she had seen in Toronto, who, like the previous two, had refused to remove her ovaries. The referral to my mentor was for the purpose of enlisting his support, as he was well known to consider it a sin to remove non-cancerous ovaries, especially prior to natural menopause. Indeed, he proselytized against this all-too-common practice that accompanied the all-too-common hysterectomies for benign uterine conditions; arguing at conferences that cancer of the ovary is rare, and ovaries are essential for the natural estrogens and other hormones required to keep a woman's bones and other organs healthy. The referral to my mentor may have also been based on the "just in case" that if ovarian cancer did develop one day, the physicians who refused to remove Miriam's ovaries would be medico-legally absolved.

I stood and spread Miriam's family tree on her bedside table, flattening out the wrinkles in front of school photographs of her two children; a daughter of about 12, and a son of about 10. I stared at Miriam's family tree. It was a sinister tree. Half of the circular leaves representing women were shaded black. Beneath each blackened circle, "ov.ca." was penciled along with a cross, indicating the death of Miriam's mother, one of her mother's two sisters, and one of Miriam's three first cousins. A second first cousin had ov.ca. under her blackened circle but no cross yet. Miriam had written the age at which each woman had died; all were in their forties, which is young for ovarian cancer. The square leaves representing the men in

Miriam's family were unshaded except for two uncles who had died of prostate cancer.[163]

I took a deep breath, staring hard at Miriam's family tree, needing to take in what I was seeing. Ovarian cancer was not supposed to be hereditary, but Miriam's family's definitely was. The heredity pattern even seemed to be autosomal dominant and of high penetrance, the most dangerous of inheritance patterns.[164] Miriam's family tree's trunk was short, having only one generation above hers. Reading my mind, Miriam sighed. "The rest of my family perished at Auschwitz."[165]

My gaze lifted to the pictures of Miriam's daughter and son on the bedside table behind her family tree. Miriam's children resembled her in their large dark eyes, almost-black hair, and high cheekbones. Their modest smiles were saying "cheese" for the school photographer. I reassured Miriam that ovarian cancer was rare and not thought to be hereditary.[166] However, I was only reassuring myself, and Miriam called me on this. I promised Miriam that she would not die from ovarian cancer.

As I was leaving Miriam's room, a particularly bright senior medical student whom we were grooming for our residency program was entering.[167] He was the clinical clerk assigned to do Miriam's pre-op history and physical. The pre-op physical did not include a pelvic exam. Because clinical clerks are graded on their

---

[163] See Chapter 18, Victor; See Chapter 25, The Arrogance of "But All You Need is a Good Index Finger."

[164] Autosomal dominant means you only need to inherit the gene from one parent to potentially develop the related condition; high penetrance means that if you do inherit the gene, you are very likely to develop the condition. Also, See Chapter 10, She Lived with the Knowledge; See Chapter 15, The Injustice of Needing Angelina Jolie; See Chapter 18, Victor.

[165] See Chapter 2, "You Must Go to Medical School or Hitler Will Have Won"; See Chapter 10, She Lived with the Knowledge; Nisker, 2012.

[166] See Chapter 18, Victor.

[167] The medical student did enter our residency program, and eventually became a gynaecologic-oncologist.

histories and physical findings,[168] they are not told anything in advance about the patient they are assigned. I turned back to Miriam. "This young man is one of our best students, and he will be very gentle." I then said to both of them, "It's okay to be shocked by what Miriam will tell you."

About an hour later, the medical student ran up to me while I was making end-of-day rounds. He was trembling. He told me he had "found something" in Miriam's left breast, "the size of a golf ball, quite hard, and with an irregular surface." He had urgency in his voice as he asked me to please come back with him to Miriam's room to "check out" what he found. As Miriam had been assessed by three physicians prior to referral to my mentor, I tried to reassure myself that the student had found a rapidly expanding breast cyst. However, his reference to an "irregular surface" caused me to walk quickly.

Miriam was sobbing and breathing quickly, overwhelmed with what the medical student might have found. She shook her head from side to side, pleading, "Please God no, please God no, please *Gotenu* no." I took Miriam's trembling right hand in mine, and asked her if I could please examine her breasts. She turned her head to the right, averting her eyes, whispered "Go ahead," and sobbed. On examining Miriam's left breast, and the lymph nodes under Miriam's left arm, it became horribly clear that Miriam would not die of ovarian cancer; Miriam would die of breast cancer.[169]

I remain riveted to this day by the injustice of Miriam's death; not only because of the injustice of cancer but because the three physicians who had assessed Miriam were too focused on her request to remove her ovaries to examine her breasts and likely uncover at an earlier stage the cancer that lurked there. I remain riveted to this day by the fear in Miriam's eyes, and her repeated words, "Please God no, please God no, please *Gotenu* no." Miriam was my first clue regarding the association of breast cancer and

---

[168] See Chapter 5, Princess Margaret.
[169] See Chapter 2, "You Must Go to Medical School or Hitler Will Have Won"; See Chapter 10, She Lived with the Knowledge; Nisker, 2012.

ovarian cancer; an association demonstrated 15 years later to be related to BRCA gene mutations.[170] However, Miriam was not my first clue that an autosomal dominant gene prevalent in women of Jewish heritage caused early onset breast cancer. Four years before I met Miriam, when I was a senior medical student on surgery at Mount Sinai Hospital in Toronto, I witnessed young women undergoing prophylactic bilateral mastectomies because of their family history of breast cancer. This strategy seemed severe to me, as at least half the women having these mastectomies would not be carrying whatever the breast cancer gene was. This strategy seemed severe to me until my Mother suffered and died from early onset breast cancer, the same breast cancer that killed her mother; the breast cancer I swore would not kill her daughter.[171]

References

Miki, Y., Swensen, J., Shattuck-Eidens, D., Futreal, P. A., Harshman, K., Tavtigian, S., Liu, Q., Cochran, C., Bennett, L. M., & Ding, W. (1994). A strong candidate for the breast and ovarian cancer susceptibility gene BRCA1. *Science, 266*(5182), 66–71.

Narod, S., Lynch, H., Conway, T., Watson, P., Feunteun, J., & Lenoir, G. (1993). Increasing incidence of breast cancer in family with BRCA1 mutation. *Lancet, 341*(8852), 1101–1102.

Narod, S. A., Feunteun, J., Lynch, H. T., Watson, P., Conway, T., Lynch, J., & Lenoir, G. M. (1991). Familial breast-ovarian cancer locus on chromosome 17q12-q23. *Lancet, 338*(8759), 82–83.

---

[170] Narod, Feunteun, Lynch, Watson, Conway, *et al.*, 1991; Narod, Lynch, Conway, Watson, Feunteun, *et al.*, 1993; Miki, Swensen, Shattuck-Eidens, Futreal, Harshman, *et al.*, 1994.

[171] Nisker, 2012, 2013.

Nisker, J. (2012). *From Calcedonies to Orchids: Plays promoting humanity in health policy.* Iguana Books.

Nisker, J. (2013). A public health education initiative for women with a family history of breast/ovarian cancer: Why did it take Angelina Jolie? *Journal of Obstetrics and Gynaecology Canada, 35*(8), 689–691.

Chapter 7

I'm Sorry Vaccine Came Too Late for You Janet

Radio warned the "snowstorm of the century" was imminent. I set my alarm for 0400 in anticipation of walking the seven kilometres to the Hospital. I trudged through the knee-deep snow in heavy darkness that was slightly softened at regular intervals by snow-burdened street lamps. With every step I thought about Janet and the complicated cancer surgery we had her "booked" for at 0800. Janet was a 29-year-old high school teacher in Northern Ontario. Janet had a large cancer of the cervix[172] that had likely been detected too late for cure. Her only hope was an onerous four-hour procedure, during which we hoped to remove her "bulky" tumour intact, leaving no cancer cells behind. A delay of Janet's surgery until the snowstorm abated would be medically complicated, and emotionally difficult for Janet.

---

[172] The cause of cancer of the cervix, the lower part of the uterus, is human papillomavirus (HPV), and the HPV vaccine has been proven to prevent cervical cancer, and is now given with permission to every Grade 7 student in Ontario (Goyette, Yen, Racovitan, Bhangu, Kothari, *et al.*, 2021). The HPV vaccine was late in being available in Canada's North, ironically the region in which cancer of the cervix is most prevalent.

I had met Janet three weeks previously, the afternoon she was transferred by ambulance from Princess Margaret Hospital in Toronto.[173] Janet was in the bed closest to the door in a two-bed room. Janet's face was emaciated from cancer's ravenous appetite. Her black-rimmed eyes had sunken into her sharp cheekbones, which were barely covered with sallow skin. Subcutaneous tissue was similarly absent from Janet's neck and shoulders. Cancer's burden inflicted an appearance reminiscent of a person in Auschwitz.[174] The rest of Janet's skeletal body was submerged under a standard nubby-textured yellow blanket. I took Janet's right hand in mine. Her bluish veins were prominent beneath her transparent skin. I introduced myself, and assured her, "We're going to help you, Janet."

After my boots plowed me to the Hospital, they climbed four flights of stairs[175] to our locker room, where I took crisply folded "greens" with blue stripes[176] on the pockets from a shelf, slipped on my waffle-sole yellow running shoes,[177] and donned a just-laundered white coat before walking over to our Ward to begin morning rounds of our cancer patients. Of course I started with Janet, who, though incredibly weak, somehow smiled at me, relieved I had made it in. She asked for reassurance that her surgery would not be cancelled. I reassured her that her surgery would be today, but likely not right at eight. Janet nodded, then smiled again as she saw my mentor[178] enter her room. Janet's smile broadened further when he lightly touched her upper left arm and gave her his warm smile. He had probably

---

[173] In 2012, Princess Margaret Hospital was renamed Princess Margaret Cancer Centre (University Health Network, 2021); See Chapter 5, Princess Margaret; See Chapter 13, The "Helix of Life" Revisited: DNA in Concrete and Not.

[174] See Chapter 2, "You Must Go to Medical School or Hitler Will Have Won"; See Chapter 10, She Lived with the Knowledge; Nisker, 2012.

[175] Claustrophobia forces me to avoid elevators. See Chapter 1, Rotor; See Chapter 3, I'm Sorry Ronnie; See Chapter 17, Ruth.

[176] Blue stripes indicated size "large."

[177] Nike Waffle-soles were wide, and thought to provide greater stability.

[178] See Chapter 6, Miriam.

decided to sleep in his office across the street from the Hospital when he heard a record-breaking snowstorm was forecast.

Janet had been assessed by three physicians in the weeks previous to her referral here, including two oncologists at Princess Margaret.[179] The radiation-oncologist felt that the bulkiness and density of Janet's cancer would render standard radiation treatment of cervical cancer ineffective, as the radiation could not penetrate deep enough through the tumour. The chemotherapist felt that no chemo had been proven effective against cervical cancer, even in reducing the size of the tumour to allow radiation to be effective. These physicians presented Janet's situation to a panel of oncologists at their weekly Case Conference, and all agreed that a "radical surgery" approach was Janet's only hope. They also agreed that the surgery would be extremely difficult, as Janet's pelvic tomogram[180] suggested her tumour might extend to her right pelvic side-wall, and if the cancer had invaded the bone, a "radical surgery" approach would be a disservice. Yet the members of the Case Conference concurred that surgery should at least be offered to Janet, and that the best surgeon to perform such complex surgery was at Western University.

In the weeks leading up to her surgery, I grew to know Janet better than any other patient during my residency; perhaps better than any patient to this day. I also grew to care for Janet more than any other patient; perhaps more than the prescribed imperative to "always be objective"[181] permitted. I ordered a series of blood tests to assess Janet's starved-status, and, ironically, had to write an "npo"[182] order on her

---

[179] See Chapter 5, Princess Margaret; See Chapter 10, She Lived with the Knowledge; See Chapter 13, The "Helix of Life" Revisited: DNA in Concrete and Not; Years later I was assessed at the new site of Princess Margaret in Toronto as described in Chapter 25, The Arrogance of "But All You Need is a Good Index Finger."

[180] Tomograms provided multi-directional images of a particular location. Tomograms were replaced by computerized axial tomography (CAT) scans.

[181] See Chapter 5, Princess Margaret; See Chapter 11, Dr. King, The Little Prince, and Seeing with One's Heart in Medicine.

[182] "Nothing by mouth" from the Latin *nil per os*.

chart, denying Janet food, even fluids except for ice chips to prep[183] Janet's bowel for surgery. This precaution was necessary to ensure Janet's bowel was as empty as possible, as we might accidentally sever her bowel in surgery or find it necessary to remove a section of bowel if it was so firmly fixed to that cancer that we could not dissect it off.

This precaution led me to the office of our Hospital's Chief Pharmacist to discuss a new intravenous-nutrition strategy called hyperalimentation[184] that I had read about in the *Canadian Journal of Surgery*. Hyperalimentation might provide Janet with the essential nutrients required to build up her body so she could tolerate the surgery, and then heal from the surgery. Twice a day, I infused the opaque-white protein and vitamin mixture through a large bore subclavian needle that I had inserted under Janet's right clavicle.[185] Twice a day I carefully checked the subclavian's insertion site to make sure it was not becoming infected, a common complication reported in the early studies of hyperalimentation.[186] I pretended nonchalance as I checked the site, engaging Janet in conversation regarding her brother and sister, what books she liked to read, what films she liked to see.

After we checked on Janet the morning of her surgery, my mentor and I began rounding on our other cancer patients. We could not help but notice that no staff seemed to be coming in. Fortunately, the

---

[183] "Prep" is short for prepare.
[184] Freeman & MacLean, 1971.
[185] I remember palpating the region in which I intended to insert the large "bore" subclavian needle, and then I drew an ink spot. Janet's eyes were watching me. I could feel her breath on the fingers of my left hand. I was amazed by Janet's lack of fear, indeed encouragement. I "painted" her skin in the region of the ink spot with Proviodine, then carefully draped a square around the spot with surgical towels. I then injected a local anaesthetic into Janet's skin so she would not feel the large "bore" needle's insertion. I then inserted the subclavian needle, drew back some venous blood, and then threaded a plastic catheter into Janet's subclavian vein. I removed the large needle, and fastened the plastic catheter onto to Janet's skin with sterile surgical tape to keep it in place. I smiled at Janet. She smiled back and said, "Thank you, Doc."
[186] Freeman & MacLean, 1971.

nurses on the night shift had unanimously agreed to remain in the Hospital to work the next day. At 0745, my mentor and I took the stairs up to our operating room to see if the anaesthetist had made it in. He had, on his snowmobile. Some of the OR nurses were beginning to come in, and were hurrying to prepare for the massive surgery that awaited their skills.

My mentor and I looked down six floors from OR 4's snowflake-draped window on the semi-circular driveway in front of the Hospital's main entrance. The ramp from the doors was being shoveled by a porter[187] whose name I no longer remember as a yellow taxi fish-tailed, struggling into the driveway. The porter opened the taxi's rear door nearest to the Hospital, as he did every morning for Doris Smith, the person in charge of the Medical Records Department. He then ran back through the Hospital's doors to where Doris's chair was waiting, and quickly wheeled it out to her. He helped Doris onto her chair and assisted her up his just-shovelled ramp into the Hospital.

As the Hospital's doors closed behind them, I heard, "Jeffer, why aren't there any of our residents in the Hospital except for you and the resident on call last night?"

I responded, "I'm not sure, sir, but I am sure they will be in soon."

My mentor was not so sure. "Jeffer, if Doris can make it in, all the residents can make it in, and I want a list of the names of the residents that didn't."

He of course knew I would not comply with his request for this list, and indeed wanted no such list; however, I ran down the steps to our floor, found the list of residents' phone numbers, and began calling. I was amazed by the phrases, "It's snowing like hell out there" and "The streets aren't plowed yet" and "It will take hours to get in." I was consistent in my response: "Start walking."

At 1015 on the day of the "snowstorm of the century," I led Janet in her hospital bed onto an elevator, pressed "6," and reassured her

---

[187] Porters are the persons who transport patients on stretchers or wheelchairs through hospital corridors, and in the past, assisted hospitals in many other ways.

calm smile with my own. However, Janet's body quivered under her just-warmed hospital blanket, betraying her all-too-justified fear. After the elevator doors opened, I flipped the keep-open switch, and led Janet's bed out of the elevator. We turned right for a few metres, and then right again for a few more metres, before turning left through the open doors of OR 4.

My mentor, the anaesthetist, and two OR nurses greeted us and helped me station Janet's bed parallel to the OR table. One of the nurses gently removed the blanket covering Janet, making sure Janet's hospital gown was pulled down to modesty. With a nurse and my mentor on one side of Janet, and the second nurse and me on the other, we grasped the bedsheet under Janet, and, with the anaesthetist steading Janet's head, gently lifted her feather-light body onto the operating table. My mentor and I looked into Janet's eyes, each of us touching one of her shoulders. She turned her head from side to side, looking both of us in the eyes, and said, "Thank you."

I smiled reassuringly, "We'll see you in the recovery room soon, Janet," before tying the white-cloth surgical mask hanging from my neck around my green-cloth surgical cap. My mentor and I watched Janet go to sleep before we went into the scrub room adjacent to OR 4 to prep our hands and arms for surgery. When we re-entered OR 4, we were gowned and gloved by the scrub nurse. My mentor then stood where he preferred, at the OR table's right side. My mentor, being a great educator, always asked the Chief Resident which side they preferred, but as I was ambidextrous,[188] it didn't matter to me. The scrub nurse handed me a sponge stick, with the gauze in its circular tweezer-clamp soaked in Proviodine. I painted Janet's abdomen from her pubic bone to just below her sternum,[189] then framed the region where we would be making the incision into Janet's

---

[188] Ambidextrous residents are confusing to the surgeons they assist, except for my mentor of course, as are left-handed residents though to a lesser degree as there are more left-handers. A surgeon once commented to me that "ambidextrous means not being good with either hand."

[189] The "sternum" is colloquially called the "breastbone."

abdomen with sterile-green towels. With the help of the scrub nurse, we covered Janet up to her neck with a sterile green sheet that had a square opening where our incision would occur. The scrub nurse handed me the scalpel.

I made a vertical incision through the skin of Janet's lower abdomen, then very slowly through Janet's almost non-existent subcutaneous tissue, and the one-cell–layer peritoneum covering Janet's intestines. Significant sections of her bowel emerged that we proceeded to pack off[190] to expose Janet's uterus, the lower half of which was enlarged with cancer. My mentor and I stared at each other as we concluded that the tumour extended all the way to Janet's right pelvic side-wall. It would be difficult surgery; surgery that would always be beyond me. I handed the scalpel to my mentor.

He carefully dissected through the layers of adhesions surrounding Janet's cancer, including the adhesions sticking Janet's bowel to her uterus. The scrub nurse and I did our best to provide exposure, using curved refractors, long tweezers, and sponge-sticks. My mentor was able to expose both of Janet's ureters, the too-often–severed tubes that transport urine from the kidneys to the bladder. He then embarked on the most difficult part of the procedure, dissecting the cancer off Janet's right pelvic side-wall. It was essential that the tumour be dissected off intact to be removed in one mass. Any spillage of tumour cells would doom Janet. After an hour of precise dissection with scalpel and fine scissors, using skill unique to him, my mentor still could not find a plane to dissect the tumour off Janet's pelvic side-wall. He began sweating profusely, something I had not previously seen. One of the nurses mopped his brow repeatedly. I knew Janet was in trouble.

After another hour of attempting to dissect the cancer free from Janet's pelvic side-wall, he was finally successful in removing Janet's cancerous uterus in one block, and transferred it into the sterile

---

[190] "Pack off" is the term we use in surgery for placing large pieces of cloth against the bowel so that we can have clear vision of the operating field.

stainless-steel bowl held out by the scrub nurse. She was surprised by its heaviness. My mentor's shoulders slumped as he stared down at a centimetre flake of cancer still attached to Janet's pelvis. He picked at it with his fine-point tweezers and scalpel, but the cancer was into Janet's bone. He exhaled deflation, turned, and walked away from the operating table, and pounded his bloody gloved fists twice against the nearest wall before bowing his head against it. I had never seen my mentor express frustration before. He rested his head against the wall for a minute, then stood erect, turned, ripped off his gloves, and threw them to the floor. He softly said, "Close, Jeffer," and left. As I sewed together the layers of Janet's incision, I felt my surgical mask moisten with tears; tears of regret for Janet.[191]

References

Freeman, J. B., & MacLean, L. D. (1971). Intravenous hyperalimentation: A review. *Canadian Journal of Surgery*, 14(3), 180–194. https://www.ncbi.nlm.nih.gov/pubmed/4997238

Goyette, A., Yen, G. P., Racovitan, V., Bhangu, P., Kothari, S., & Franco, E. L. (2021). Evolution of public health human papillomavirus immunization programs in Canada. *Current Oncology*, 28(1), 991–1007.

Nisker, J. (2012). *From Calcedonies to Orchids: Plays promoting humanity in health policy*. Iguana Books.

University Health Network. (2021). Princess Margaret history. https://www.uhn.ca/corporate/AboutUHN/OurHistory/Pages/princess_margaret_history.aspx

---

[191] When I finished "closing" Janet's abdomen, I stared at her intubated face, consumed by the injustice that cancer had inflicted upon her. It would be up to me to tell Janet that we were unable to remove all the cancer, and answer her direct question regarding her prognosis. It would be up to me to say, "I'm sorry Janet."

Chapter 8

Thank You Grace

It was my first weekend on call
As a certified rookie specialist
Fresh from three extra years
Of a cancer-research fellowship
It was also the weekend of our Annual National Conference
And not surprisingly I had won the lottery
To cover for my senior colleagues.

On Saturday morning I was paged
To take a call from an "upcountry" ambulance,
Requesting I accept a woman
Hemorrhaging from her sixth birth
Two hours later the ambulance warned
To expect the woman in ten minutes,
And that her vitals were so unstable
She probably wouldn't make it.
I crashed down four flights of steps
To Emerg where out of breath
I expected death as did the nurses
Who preferred death's declare before their door.

The back of the ambulance was quickly opened
And Grace's stretcher rushed through the ER doors,
Where her light blue eyes greeted me,
And her serene smile smoothed me in peace.

Grace was colloquially "white as the sheets,"
Having no blood extra to breathe her skin,
But her cognition was crystal clear,
And her smile beyond engaging.

I touched Grace's left wrist
To introduced myself and take her pulse,
Which was less rapid than her blood loss anticipated
And more regular than my own.

I asked Grace if I might examine her,
And assured "I'll be gentle"
Her smile concurred as her words whispered,
"Please go ahead."

I placed my palm on Grace's abdomen.
And found her uterus was up to her ribs,
Likely filled with cloth packing
That inhibited her uterus from contracting.

Her uterus must have been atonic,[192]
As uteruses become after many births,
And the packing's tamponade of her bleeding,
Kept Grace breathing but needed removing.

---

[192] An atonic uterus lacks the ability for its muscles to contract down and compress the blood vessels bleeding on the placenta's implantation site after the placenta has been removed.

Grace had four IVs running
Trying to maintain blood-vessel volume,
But none of the intravenous contained blood,
Because of Grace's firm Faith.

Blood-substitute products were not yet available
So pushing an intravenous antibiotic-rich solution
Was the only solution available to Grace.
Even though the solution did not include hemoglobin's oxygen.

Grace's minimal hemoglobin
Would dip even lower
When diluted by "open" IV fluids,
Indeed, set a Hospital record

Grace could not afford to lose more blood,
But I knew she would when I removed her pack,
A procedure essential to ensure
The pack's accelerating bacteria would not spread Grace's death.

Bacteria is the reason
We do not pack a uterus,
But the "upcountry" physician did,
And his pack saved Grace.

Just after my assess of Grace,
I pushed her stretcher straight to Intensive Care,
Where its physician was reluctant to accept her
After he learned Grace would not accept blood.

Indeed he was incensed with the possibility
Of a preventable death in his Unit
Taxing his resources and statistics
And disturbing his nurses.

However my obligation to Grace
Firmed my resolve that all that could
Would be done for Grace
Starting with Intensive Care admission.

I dipped the ICU bed's head
Into Trendelenburg position
To encourage Grace's remaining blood
To flow from her legs into her head.

The nurses further encouraged this
By wrapping Grace's legs in inflatable stockings,
Then when dilated would diminish
Hemoglobin's oxygen pooling in this region

I was in constant friction with the ICU physician
Because he kept battering Grace about blood transfusions,
Under the illusion his power would brow-beat her
Into submission despite Grace's devotion.

The ICU tried to soften to sleep,
Amidst consistent bellows-breathing and beeps,
And nurses curtained suctions.
And fluorescent-light interruption.

Grace further consented her understanding
To a procedure that could not be avoided
And made me promise to never condemn myself for co-operating
    with her decision.

Through the night I sat with Grace,
Under the spectre of pulling the pack in the morn,
And facing the hemorrhage that would surely flow
And my impotence to replace low hemoglobin.

Grace explained to me softly
That I should not fear her death,
Or mourn the firmness of her decision
For she would soon "sit with Jesus"
And I should be happy for her.

Grace's family members began congregating
Along with her pastor and members of their faith-community,
In the ICU's waiting room,
One at a time permitted in the Unit.

However at dawn the ICU physician
Assembled them all around Grace's bed,
And said "Grace say goodbye to your family,
Because when Dr. Nisker pulls your pack you'll be dead."

It was a horrible assault on Grace,
But she did not flinch nor did her family members,
Nor were serene expressions altered,
Because death was not an end but a beginning.

But I did flinch and Grace stared
At the upset in my eyes,
Stemming from the ICU physician's
Condemning of Grace.

Then Grace smiled a reassuring smile,
Seeming more concerned with me than with herself,
And reached out her intravenous left hand
To calm the quiver in mine.

I grew to know Grace that night
And understand her concern for me,
Because she did not want to inflict her death
On a person who did not accept Jesus.

I reflected as I touched
Grace's hand till we went to the OR
Reflected about Grace's
Touching concern for me.

In the OR before 0800
The anaesthetist shouted
It was his right not to participate
Because anaesthetic agents would encourage
Blood-vessel dilation and further hemorrhage.

And he reminded me that my patient's
Hemoglobin was just 2.4,
A probable record for a conscious patient,
And he was amazed she was still alive.

Two nurses and I lifted Grace
Onto the operating table
And tilted the head of bed
Down as far as we were able.

Then with Grace's permission I lifted her legs into the stirrups,
A hereditary procedure
I would soon refuse to go along with.

I gently placed a speculum in Grace's vagina
And observed the blue string indicating the end of the packing,
Grasped it with a sponge stick and slowly began pulling.

Grace and I conversed continuously
As I urged the long chain of pads from her uterus,
All the pads were soaked dark red
With old blood from her hemorrhage.

The chain of packing became three metres long,
But I was heartened to notice
No evidence of bright red
That would suggest active bleeding.

I kept sharing this finding with Grace,
Who kept encouraging me with her smile,
For there would be no need for the hysterectomy
For which signed in case of hemorrhage.

Grace's "upcountry" physician's
Misguided decision to pack her uterus
Had saved Grace's life,
And given me a great gift.

Chapter 9

Beneath the Pagoda's Perch

On the highway to a Narrative Ethics conference in Ohio,[193] I was transfixed by a sign indicating the "Town of Kent." As soon as I arrived at the conference, I began asking participants to join me on a pilgrimage to Kent State University. A woman about my age offered to go with me, quietly adding she had been a student at Kent State in 1970 and had witnessed the tragedy of May 4.[194] The next day she led me to a hill, still topped with the indelible pagoda-like rain shelter engraved in my brain. I knelt at the pagoda and looked down on the field where the National Guard had fired 70 rounds into unarmed students peacefully protesting the Vietnam War, killing four, and injuring many others.[195] I was barraged by afterimages of newspaper photographs and evening news footage.

Although I was a Canadian medical student, safely snug from the Vietnam War above a border an hour and a half south, I grew up watching American television, and saw no distinction between myself

---

[193] "Ohio" is the title of a song, commemorating the "four dead in Ohio," written and composed by Neil Young and performed by Crosby, Stills, Nash & Young (1970).
[194] Gordon, 1990.
[195] Gordon, 1990.

and American students. So when lots were drawn from a fishbowl to condemn teenagers exactly my age to the Vietnam War, I shared their sense of betrayal, as I had shared their sense of betrayal two years previously when Dr. Martin Luther King Jr.'s life was extinguished.[196] I share their betrayal each time I kneel before the names engraved in the black-marble wall extending from President Lincoln's feet, where Dr. Martin Luther King Jr. gave his "I Have a Dream" speech to believers at the culmination of the March on Washington.[197]

I wrote this poem not only to commemorate the students killed and injured at Kent State, but to commemorate the day I first anticipated the world might not be changed by my generation's enlightened efforts, the day I first felt the foreboding that the next generation of students might not live in the utopia my generation had imagined; the day my youthful optimism perished in adult reality. "Beneath the Pagoda's Perch" was posted on the Kent State In-Memorial Wall, and was included in the Jerry M. Lewis papers May 4 Collection in Kent State University Library.[198] On May 4, 2020, the 50th anniversary remembrance ceremony of the Kent State massacre had to be cancelled due to COVID-19. I spent the day reflecting on how so little had changed, particularly regarding the murders at the hands of men of authority,[199] including those of Sandra Bland, Eric Garner, Trayvon Martin, Tamir Rice, and George Floyd.[200]

---

[196] King, 1986.
[197] Gordon, 1990; History.com, 2010.
[198] Nisker, 1995.
[199] Brown, 2019.
[200] "14 high-profile police-related deaths of U.S. Blacks", 2017; Brown, 2019; "George Floyd told 'it takes … a lot of oxygen to talk' during arrest, transcript reveals", 2020.

References

(2017, Dec 7). *14 high-profile police-related deaths of U.S. Blacks.* Canadian Broadcasting Corporation News. https://www.cbc.ca/news/world/list-police-related-deaths-usa-1.4438618

(2020, Jul 8). George Floyd told "it takes … a lot of oxygen to talk" during arrest, transcript reveals. Canadian Broadcasting Corporation News. https://www.cbc.ca/news/world/george-floyd-chauvin-arrest-transcripts-1.5642789

Brown, J. (2019). Bullet points. In *The tradition* (pp. 16–17). Picador. https://www.poetryfoundation.org/poems/152728/bullet-points

Crosby, Stills, Nash, & Young. (1970). Ohio [Song]. Atlantic.

Gordon, W. A. (1990). *The fourth of May: Killings and coverups at Kent State.* Prometheus Books.

History.com. (2010). *Martin Luther King, Jr. delivers "I have a dream" speech at the March on Washington.* This Day in History. A&E Television Networks. https://www.history.com/this-day-in-history/king-speaks-to-march-on-washington

King, M. L. (1986). *A testament of hope: The essential writings and speeches.* HarperOne.

Nisker, J. (1995). Beneath the pagoda's perch: A Canadian at Kent State. In *Poetry about Kent State shootings.* Kent State Shootings: Digital Archive. https://omeka.library.kent.edu/special-collections/items/show/4448

Beneath the Pagoda's Perch

A quarter century since a teen turned home,
She leads me to a tapestry not afforded the respect
Of older battlefields where America shot their own
And sees a day we must never forget.

With wrath wrapped warmly in lips quivering care,
Tortured trust spills through staccato sighs,
As her shoulders swallow apostasy's grey hair,
And hurt exhales through imploring hazel eyes.

Her soundless syllables seek the apologue's soul,
Hands that wove healing still question cause,
As her heart unravels its woolly roll,
Compassion conjures as I gaze through grief's gauze:

At photos fluoresced down AP chains,
At anchormen, solemn affect attired,
At grotesque engraving on my vision's veins,
At cacophony tattooed by sincerity satired;

At students decrying Cambodia's invasion,
Pelting youth's words against a tear gas wall,
Doomed to decay by fears' abrasion,
Bravely bear-baiting until fantasy's fall;

At uniformed pawns ill-prepared for the game,
Students' age but dealt poorer cards,
Kneeling the sentence of society insane,
Real bullets filled the toys of the National Guards;

At the pagoda preaching from higher,
Surprising the supplicants to absurdity's sport,
Our purity's epiphany consumed on its pyre,
That witnessed the waste of the rifles' report;

At the warheads' wanton smoke-stained sky,
Whispering random ravages despair;
At two women, two men, protesting, passing by,
As futility's fumes fouled Ohio air;

At candlelight proclaiming innocence incensed,
At placards pleading responsibility discover,
At the intolerance ingrained when "control" commenced,
At our youth failing and never to recover;

At the sorrows of '68 still grieved,
The dreams' extermination just recently pried loose,
Halted by horror too profound to be perceived,
Naiveté reduced in reality's noose;

At myself beyond the capture of the fishbowl's fate,
Even though damned to death's deck by my '49 birthdate,
I shared my southern cousins' pain,
I mourned and marched but in safety remained.

Her skein exhausted, warp's sapience stowed,
Her passion worn fairly through the sanguine search
Of adolescence betrayed by the behemoth's blow,
That buried Pollyanna beneath the pagoda's perch.

Chapter 10

She Lived with the Knowledge

She lived with the knowledge it would happen to her
Knowledge more felt than understood
Knowledge gleaned from intuition that could not be confessed
Knowledge that always lived and never rests.
She lived with the knowledge it would happen to her
Woke each day to the knowledge it would happen to her
That what happened to her Mother would happen to her
She only wondered when it would happen and when would it end.

My knowledge began with my Grandmother
With whom I lived at my life's beginning
Who lived with us at her life's ending
My wonderful second mother.
Powdered in kitchen table flour
She told stories of Schweitzer and Hammarskjöld
Patiently engraving her goodness
To proxy me with their purpose.

Breast cancer found my Grandmother when she was forty-four
I was sixteen when she died a few years later
I did not know my Grandmother had breast cancer
I did not know she would soon die.
Though she suffered surgery and chemotherapy
Radiation and fluid taps
They were carefully hidden behind parental backs
Forbidden to my adolescent distraction.

But perhaps those backs were less opaque
And it was I who willingly chose to take
Each molecule of selfish density to deny
Truth to a teenager too enamoured with teenage rise.
Even when my Grandmother moved in with us
And lived in a hospital bed in my room
I forbid contemplation of where that bed led
Till my Grandmother was led to a hospice-hospital.

A few days later my Mother's telephone-whisper
"Jeff come to the hospital, but don't tell your Brother or Sister"
Insisted it would be the last time
My Grandmother's eyes and mine would entwine.
Yet as I held her goodness in my hand
It was the first time our eyes entwined in her truth
We stared hard at each other
Long after the "Visitors' Hours are over."

I could not contemplate leaving her
Because she would never leave me
As long as I held the hand that held me
The hand that still holds me.
I wore her colours in this joust with injustice
And would not be unseated
Till "too young" rules decreed I must get up
And my trust of justice was lanced forever.

I loved my Grandmother very much
For years I daily grieved
But never for one moment perceived
That what happened to my Grandmother would happen to her.
Seven years later medical print premonisced
My Mother would suffer the same injustice
The fact soon fell from its suspended shelf
When my Mother discovered her assassin herself.

It had been only two months since
The mammogram I arranged proclaimed "all clear"
My Mother bravely bore her long-accepted bier
And missioned to soften her family's fear.
Mastectomy delivered a tiny stone
The surgeon delivered an optimistic poem
"No spread, no further treatment to tread"
But the pathology report delivered an "aggressive" retort.

This surgeon shared "aggressive" with just the physician-son
As it was a time when cancer-patient families
Were encouraged to cheery possibilities
Through tones that knell "All's well."
The physician to physician share
Of all may not be well
Was a care I reluctantly did not share
With my pleasantly pastelled family.

The tumour's small size and negative nodes
Bode no tamoxifen, no radiation, no chemo
Tamoxifen was new then and thought to later lend
Leukemia to women it borrowed from breast cancer.
Radiation and chemo would hurl her further abuse
I could not advocate their adjuvant use
Not when the sure surgeon's words of optimism
So soothed my reassured family.

I abrogated "aggressive" awareness
And acquiesced my Mother to another's care
I let the surgeon take command, and went to California
On a breast-cancer research fellowship.
Of course I now admonish this acquiescence
I should have spent her each remaining day
Finding ways to repay the love she lavished
But my Mother would not hear of it.

A year later my Father's long-distance words
"Jeff I have some bad news"
Collapsed my knees, my lungs, my life
As it confessed the poison she possessed.
Nothing further needed to be said
I knew my Mother would soon be dead
My silence heard my Father urge chemo would cure her
A reassurance irrelevant to my Mother's reality.

I flew home viewing 8 mm movies of my Mother
There were so many smiles in so few years
No sound was needed to hear her embrace
No colour required to feel her grace.
My friends likened my Mother to Maria in *West Side Story*
But I view my Mother as more beautiful
Eyes smiling on a dance floor as she spins beneath my arm
Eyes in love as she wraps herself in my Father's arms.

I hurried the hospital's revolving door
Then sped the elevator to the cancer floor
Where its doors opened on a black "In Loving Memory"
Engraving the names of the cancer-killed.
I quickly glanced at my Grandmother's golden name
Avoiding noticing the black space below
That would soon be embossed with the engraving
Of my Mother's so golden name.

I ran fast to the nurses' station
Breathlessly begging my Mother's room number
I ran faster to that room and parted the curtains
Only to find a woman who was not my Mother.
The woman smiled at me warmly
Her head bare in a wheelchair
I said "Sorry" and bolted to the room next door
Before being locked in the abhor that I had just spoken to my
    Mother.

Panic punished as I tried to undo my betrayal
I ran back to her room to her unvanquished smile
To her "Don't worry you didn't recognize me without my hair"
I hugged her waist with my head and begged her forgiveness.
My Mother locked her fingers in the curls of my hair
And quickly released me to her comfort
My sobs flooded my Mother's gown
Where I felt I had drowned.

With each week's advance my family firmed their faith
That more chemo would turn metastatic cancer's advance
Sympathetic nurses encouraged this not-credible credulity
I of course knew no such even-temporary luxury.
I silently shouldered death's imminent answer
As cancer poured its spores through my Mother
Growing tubes to remove its seditious sap
From her abdomen, her bladder, her brain, her back.

Soon the cancer that seeded her brain
Convulsed her diminishing body
The doctors countered with their sweet sedation
That took her from us before we were ready.
I asked the drugs be backed off to bring my Mother back
And her crystal comburence warmed another month
Till rare became the moments my Mother was aware
Of the love that surrounded her, the love she put there.

My family denied my Mother's imminent death
Constantly advising of the latest newspaper finds
Such as Laetrile and other unevidenced medicines
That were popular in the press at that time.
My gentle urge that there was no magic cure
In the world that could miracle my Mother
Was met with contempt and haranguing
That my heart had been hardened in medical training.

The elevator doors opened for the last time
The "In Loving Memory" list embossed my imminent loss
My eyes tunnelled to resist my Grandmother's golden name
And the space where my Mother's name would soon embrace hers.
My Grandfather met me near the elevator to eagerly effuse
"There's a new drug on the news that is sure to cure your Mother"
I hugged his helplessness and walked him to my Uncle and Aunt
Pacing the verandah to the vigil I would enter for the last time.

As my Mother began her terminal-breathing pattern
My Sister wept on my Mother's chest
My Father sat in a chair staring at his forever love
Glad her suffering would soon be over.
My Brother and I stood bookending her bed
Staring down at the kerchiefed head we so loved
Silent centurions guarding the gates of the dead
Steadfast in prohibiting our Mother's passage.

As I took my Mother's wedding-ringed left hand
My Brother took her right hand
We were touching an angel's gossamer wings
As they slowly spread for flight.
When her final breath exhaled her death
My Brother commenced cardiac massage
I refrained his wrists in whispered gauze
"She's gone but she will never leave us."

His eyes glazed over but mine could not
I lifted them to my Father's searching mine for what he already knew
We stared at each other for more than a long time
Longing for her, allies in loss.
My eyes turned back to Mother's eyes
Closed in peaceful sleep for eternity
I held my Mother's hand past cold
I still hold her hand and will forever.

I love my Grandmother very much
I love my Mother very much
I love my Sister very much
Who lives with the knowledge it will happen to her,
Her knowledge is more felt than understood
Her knowledge gleaned from intuition that cannot be confessed
Her knowledge that always lived
Her knowledge that it will never rest.

She lives with the knowledge it will happen to her
Wakes each day to the knowledge it will happen to her
That what happened to her Grandmother will happen to her
That what happened to her Mother will happen to her.
She wonders when it will happen
She wonders when it will end
But "it" will not happen to my Sister
"It" will not happen to her daughters.[201]

---

[201] To prevent "it" from happening to sisters and daughters in other families, I wrote *Sarah's Daughters* for public education regarding BRCA gene mutation breast cancer in the late 1990s. "She Lived with the Knowledge" had been just written in response to the request of Cancer Care Ontario to write a poem to be read at the opening ceremonies of its annual conference. *Sarah's Daughters* is a 90-minute play incorporating scientific information about BRCA gene mutations, and about the prevention, early diagnosis, and treatment of BRCA-related breast and ovarian cancer. I morphed "She Lived with the Knowledge" into Act I Scene 1 of *Sarah's Daughters*. *Sarah's Daughters* toured Canada multiple times, as well as touring four other countries. It was performed for thousands of women, many of whom likely had a high hereditary risk of early onset breast cancer and were attracted by the theatre posters and

other advertisement indicating the subject matter of the play. *Sarah's Daughters* drew on events that occurred in Canada regarding the lack of availability of BRCA gene mutation testing, which remained confined to a research protocol long after it was clinically available in other countries. Further, the research protocol ignored the fact that the highest risk group, Jewish women whose ancestry was from Eastern Europe, were less likely to have an adequate number of women with breast cancer in their surviving family members to qualify them for testing. By focusing on the story of one woman, her family, and her best friend, the play also aimed to place the ethical, health policy, and scientific issues in a context that was accessible and engaging for the general public and health-policy makers.

Chapter 11

Dr. King, The Little Prince, and Seeing with One's Heart in Medicine

It is only with the heart that one can see rightly; what is essential is invisible to the eye.[202]

—Antoine de Saint-Exupéry, *The Little Prince*

Every evening with brushed teeth, my pajamaed children huddled with me in the hall between their bedrooms beneath a poster of Dr. Martin Luther King Jr. and his dreams.[203] Dr. King's compassionate

---

[202] As claimed by the wise Fox in *The Little Prince* (de Saint-Exupéry, 1943).
[203] "I have a dream that one day on the red hills of Georgia the sons of former slaves and the sons of former slave owners will be able to sit down together at the table of brotherhood. I have a dream that one day even the state of Mississippi, a state sweltering with the heat of injustice, sweltering with the heat of oppression, will be transformed into an oasis of freedom and justice. I have a dream that my four little children will one day live in a nation where they will not be judged by the color of their skin but by the content of their character.

eyes gazed down on us, just as they gazed down on the hundreds of thousands hearing his dreams at the culmination of their March on Washington in 1963;[204] a march to encourage Congress to pass President Kennedy's Civil Rights Bill. Dr. King met with President Kennedy immediately after his speech to also seek a second piece of legislation, the Voting Rights Act. President Kennedy was assassinated a few months later, so that act had to wait two years to be signed into law. Dr. King was assassinated on the eve of yet another of his social-justice marches a few years later.[205]

Beneath Dr. King's compassionate eyes, we sang the social justice songs of Bob Dylan and Joan Baez, Buffy Sainte-Marie and Pete Seeger. Beneath Dr. King's compassionate eyes, we read the social justice writings of Gandhi and Lao Tzu, Karl Marx and Confucius. Beneath Dr. King's compassionate eyes, we read parts of his social-justice speeches. We called these intimate moments our "Thoughts for the Night," a termed coined by a frequent sleepover-friend of one of my sons. Although the Thought for the Night seeded social justice in the dreams of my children, it also seeded sleep-deprivation discussions that would reawaken on long car rides, forest walks, and canoe paddles. My strategy for imbuing social justice in my children was different from my Grandmother's strategy for imbuing social justice in me, but not by much.[206] My strategy for imbuing social

---

I have a dream today. I have a dream that one day down in Alabama, with its vicious racists, with its governor having his lips dripping with the words of interposition and nullification, that one day right down in Alabama little black boys and black girls will be able to join hands with little white boys and white girls as sisters and brothers. I have a dream today" (King, 1986).

[204] Euchnere, 2010.

[205] Many other American civil rights leaders were also killed by assassins' bullets, including the president's brother, Robert Kennedy, two months later; and other civil rights leaders were lynched from trees or burned in churches by white-hooded Klansmen.

[206] See Chapter 2, "You Must Go to Medical School or Hitler Will Have Won"; See Chapter 5, Princess Margaret; See Chapter 10, She Lived with the

justice in my children became my strategy for imbuing social justice in medical students and physicians.

Our University's Westminster Institute of Ethics was an early casualty of our new tax-break–elected provincial government's funding cuts to higher education and hospitals, as it was funded in half by both. Women's shelters and affordable daycare facilities were also early casualties of funding cuts. Another early casualty was the provincial-police sniper's killing of Dudley George, an Indigenous man participating in a peaceful land-ownership protest at Ipperwash Provincial Park, 45 minutes from our home.[207] During the subsequent inquiry, the premier's testimony acknowledged that he wanted the occupation brought to a quick end, but he denied he had said "I want the fucking Indians out of the park,"[208] although this command was heard by many. Following the police-assassination of Dudley George, I was recruited by Rev. Susan Eagle[209] to be her "white-male-physician" spokesperson, whose voice she believed was more likely to be heard by the men in power than hers; particularly in regard to exposing the injustices inflicted on socio-economically disadvantaged women.

Following the withdrawal of funding from our Institute of Ethics, several of its philosophers left our University, including the philosopher who taught bioethics in our Medical School. The Dean of Medicine asked me to take over the six-hour "fluff-course," likely thinking I was capable of teaching bioethics because of the predominant assumption among physicians that any physician can teach bioethics because we all practise in an ethical manner. My Dean's thought process in choosing me was likely also based on the fact that the premier's cutbacks to my Dean's budget insisted

---

Knowledge; See Chapter 13, The "Helix of Life" Revisited: DNA in Concrete and Not.

[207] Dubinski, 2020.

[208] "Ipperwash inquiry spreads blame for George's death," 2007.

[209] Rev. Susan Eagle famously cream-pied the premier a few weeks later on the steps of the Provincial Legislature's building.

someone already drawing a salary from the Faculty of Medicine take over the course. My Dean's thought process may have been further encouraged by my lengthy submissions to our University's Human Research Ethics Board that he had to sign, if not read.[210] I did know enough about bioethics to know I knew very little about bioethics; certainly not nearly enough to teach it.

With my knowledge of my lack of knowledge of bioethics, I responded to my Dean with, "I'm but sorry I'm not qualified to teach bioethics." The Dean was of course disturbed that a request he made to one of his faculty was not met instantly with an unequivocal "Yes." He furrowed his brow at me, so I quickly modified my response to the man who paid my salary with, "Let me think about it for a bit."

While thinking about it for more than a bit, my brainwashing of my children to social justice through our Thoughts for the Night was working so well that I shamelessly shared it with colleagues over the closing dinner of a national medical conference. The colleague sitting next to me asked if I had ever read my children *The Little Prince*.[211] I had not. Indeed, I had no knowledge of the existence of this book. The following morning while waiting for my flight home, I was surprised to see four copies of *The Little Prince* on a shelf in the Ottawa International Airport bookstore, and of course bought one. I shoved the thin *The Little Prince* into my stuffed-to-capacity briefcase, and instantly forgot about it until a line in the poem we were reading during a Thought for the Night the following week tweaked my memory. I excavated *The Little Prince* from my briefcase.

The story of *The Little Prince* was unusual, even for a children's book, as were the simple illustrations Antoine de Saint-Exupéry had painted. I hesitantly read the book to my children, who were absorbed

---

[210] Perhaps my Dean also thought me capable because of having become an "associate" of our Institute of Ethics just prior to its closure because of my frequent road-crossings to consult with one of its philosophers regarding my proposed reproductive-genetics research that was complex from both an ethics and scientific perspective.

[211] de Saint-Exupéry, 1943.

by the story, and even more by the paintings. *The Little Prince* was not like anything we had read; indeed, would ever read. Although my children seemed enthralled by the story and paintings, I was lukewarm about both until I read the wise Fox's words, "It is only with the heart that one can see rightly; what is essential is invisible to the eye."[212] I must have reread this line several times because my eldest asked in a concerned voice, "Are you okay, Dad?" I promptly responded, "It's time to go to bed." I knew I had read something epiphanic and needed time to reflect on its meaning.

That night I repeated "It is only with the heart that one can see rightly; what is essential is invisible to the eye" over and over. Suddenly with crystal clarity, I realized seeing "with the heart" was "essential" in protecting the hearts of medical students from succumbing to the numbing of medical education's objectification of persons; objectification that diminishes medical students' capacity to become compassionate physicians. This objectification had been purposefully designed for multiple-choice–efficient computer-evaluation formats; formats that distract from in-depth examination of the complexities that can make us compassionate physicians. The wise Fox's words brought me to the understanding that it was the compassion in the eyes of Dr. Martin Luther King Jr., emanating from the compassion in his heart, that made Dr. King's words so compelling.

The wise Fox's words echoed through my mind day and night during the weeks before my Dean next called me into his office to insist I take over the bioethics course. With the wise Fox's words embedded in my brain, I responded to my Dean with a "Yes but," as in "Yes, but only if I can use story to engage medical students' hearts as well as their minds in bioethics." I quickly added that I would need more curricular hours to accomplish this. I defended the increase in hours by explaining that I planned to abandon the abbreviated "Bioethics Principles" approach to bioethics pervasive in medical

---

[212] de Saint-Exupéry, 1943.

schools because of the first edition of *Principles and Theories of Bioethics*.[213] This "bioethics bible" was written specifically for medical students by two American philosophers, Tom Beauchamp and James Childress, based on the well-known assumption among philosophers that medical students are not truly intelligent, just good at guessing the answers to multiple-choice questions; after all, that's how they got into medical school.[214] I was convinced most medical students would be receptive to seeing with their hearts and eventually privilege their hearts over the medical texts inscribed in their brains, and thus become the compassionate physicians I was sure they wanted to be. I kept seeing the compassionate eyes of Dr. King.

The increase in hours requested for our "Narrative Ethics" program could of course not be found in the core curriculum, so I was offered the extra hours on Monday evenings. The students termed our ethics and social-justice program "Monday Night Narratives,"[215] and in 1996 the seeing-with-one's-heart students took on an intense lobbying campaign to persuade our Dean that our Narrative Ethics Program deserved core-curricular attention. Medical-student lobbying eventually expanded the program into a compulsory 200-hour core curriculum course.[216]

Each three-hour ethics exploration began with a "Readers' Theater"[217] of a short story, poem, scene of play, speech by Dr. King, or the group-singing of a social-justice song. I projected "all of the

---

[213] Beauchamp & Childress, 1979.

[214] Many medical schools have gradually deemphasized the importance of MCATs and other multiple-choice entrance exams; instead, privileging interviews and essays in medical student selection. However, interviews introduced racial bias that computers did not have. Many of the best medical schools still privilege some form of multiple-choice, computer-scored entrance exams.

[215] Nisker, 1997.

[216] Nisker, 1997, 2004, 2010.

[217] I first observed the concept of "Readers' theater" at the Center for Literature and Medicine in Hiram, Ohio, in 1996.

above"[218] on a screen covering the chalkboards. An acceleration in class attendance soon followed, as did compassion in medical-student eyes instead of the standard "ethics-eyelid reflex."[219] I sensed empathy in the medical students' now compassionate eyes, and passionate questions; empathy with the persons in the narratives submerged in the vortex of health problems. It was time to take the wise Fox's wisdom to practising physicians,[220] to health-policy makers, and to the general public[221] by writing full-length plays,[222] and eventually the novel *Patiently Waiting For*....[223] These plays and my novel were written with the compassionate eyes of Dr. King looking over my left shoulder.

I had needed the wise Fox's words,[224] as read to my children beneath the compassionate eyes of Dr. Martin Luther King Jr., to have insight into how to engage the hearts of medical students, physicians, and other health professionals in compassionate healthcare practice. I had needed the Fox's words written in 1943 to promote social justice in 21st-century health-policy development. I had needed Dr. King's compassionate eyes looking over my shoulder to be a better physician, better educator, better father, better person.

I pay tribute to Dr. King on Martin Luther King Jr. Day, the third Monday of January every year. I try to pay tribute to Dr. King every day, in what I say, in what I write, and in how I behave.

---

[218] "All of the above" refers to the bottom choice on many multiple-choice questions.
[219] Nisker, 1997, 2004, 2010.
[220] I bordered on being evangelistic in my belief that seeing with one's heart was essential in our clinical practice, and even encouraged the literature of fiction to be read by alongside the literature of medicine.
[221] The general public, through their votes, are also health-policy makers.
[222] Nisker, 2012.
[223] Nisker, 2015.
[224] de Saint-Exupéry, 1943.

References

(2007, May 31). *Ipperwash inquiry spreads blame for George's death.* Canadian Broadcasting Corporation News. https://www.cbc.ca/news/canada/ipperwash-inquiry-spreads-blame-for-george-s-death-1.666937Beauchamp, T. L., & Childress, J. F. (1979). *Principles and theories of bioethics.* Oxford University Press.

de Saint-Exupéry, A. (1943). *The little prince.* Harcourt Brace Jovanovich.

Dubinski, K. (2020). 25 years after his death, Dudley George's fight for the land continues. Canadian Broadcasting Corporation News. https://www.cbc.ca/news/canada/london/dudley-george-kettle-stony-point-25-anniversary-1.5708055

Euchnere, C. (2010). *Nobody turn me around: A people's history of the 1963 March on Washington.* Beacon Press.

King, M. L. (1986). *A testament of hope: The essential writings and speeches.* HarperOne.

Nisker, J. (2010). Theatre and research in the reproductive sciences. *Journal of Medical Humanities, 31*(1), 81–90.

Nisker, J. (2012). *From Calcedonies to Orchids: Plays promoting humanity in health policy.* Iguana Books.

Nisker, J. (2015). *Patiently waiting for...* Iguana Books.

Nisker, J. A. (1997). The yellow brick road of medical education. *Canadian Medical Association Journal, 156*(5), 689–691. https://www.ncbi.nlm.nih.gov/pubmed/9068580

Nisker, J. A. (2004). Narrative ethics in health care. In J. Storch, P. Rodney, & R. Starzomski (Eds.), *Toward a moral horizon* (pp. 285–309). Pearson Education Canada Inc.

Chapter 12

For Medical Students Protesting the Injustice of Clayoquot Sound

In 1993 a group of medical students was heading to Clayoquot Sound to protest the injustice of the clear-cutting of the region's old-growth forests that we were witnessing on television news programs. The First Nations peoples of Clayoquot Sound saw the clear-cutting as beyond injustice; it was sacrilegious. The students asked me to go with them, but I had to decline because of clinical responsibilities. In a feeble instead, I wrote a protest song for the students to sing while chained to the trees, while being arrested, and while in jail cells. I also promised I would get them out of jail. My musician-son contributed music to my protest lyrics.

In 2018, illness finally gifted me a visit to Clayoquot Sound. As I flew in, the enormous bare-square scars that frequented British Columbia's forest brought me down before the plane descended. I had never been to such a spiritual place. Following a Tla-o-qui-aht guide through the old-growth forests, I basked in the sacredness. I had never been to such a spiritual place. I could not help but contemplate why the enormous square-desecrations of lumber companies covering British Columbia had been permitted to happen.

The inspiration to write these lyrics is one of the many gifts given to me by my students.

Clayoquot Sound

Verse 1
: Through the window of a jet plane
Looked down and found no forest remained.
I asked the name of this tragic ground
Someone whispered "Clayoquot Sound."

Verse 2
: Where leviathans once touched the sky,
Till the government sold their sentence to die.
The trees grew bold a thousand years I was told,
Till greed reduced them to newspaper rolls.

Chorus
: We'll never ride that undertow again
Never provide a dead-end road again my friend,
Plant understanding signs along the way
Make compassion happen every day.

Verse 3
: To keep Truth hidden from our evangelist eyes,
The forest industry, their desecration disguise,
By leaving fringes of forest around,
Highways, rivers, lakes, and towns.

Verse 4
: But from the air, bare squares are seen,
Witness to where clear-cut carnage has been;
Sacrilegious scars on divine creation,
Defying the doctrine of reforestation.

Chorus	We'll never ride that undertow again Never provide a dead-end road again my friend, Plant understanding signs along the way Make compassion happen every day.
Break	Everywhere the rage repeats: Forests buried 'neath bulldoz'n' feet, Every tree is cost-effectively killed. To provide their bodies for the lumber mills.
Verse 5	The native people now banished from their lands, Their gardens of glory fed consumer demands. When ideals are chopped for profit, Nothing remains of beauty that was Clayoquot.
Chorus	We'll never ride that undertow again Never provide a dead-end road again my friend, Plant understanding signs along the way Make compassion happen every day. Make compassion happen every day.

Chapter 13

The "Helix of Life" Revisited: DNA in Concrete and Not

I returned to my all-degree alma mater in 2017, and stood beneath the "Helix of Life." The towering concrete twist of a DNA double helix, planted at the main entrance to the Medical School's brand-new building when I was a brand-new medical student, was meant to represent the entrance to the future of medicine through new genetic technologies. The Helix would soon be the site of my first social-justice protest as a medical student. In 2017, I stood beneath the Helix just as Canada's *Genetic Non-Discrimination Act*[225] was set to be passed by Parliament to prohibit genetic discrimination in securing life insurance and employment; however, I was upset that the *Act* failed to address the injustice of systemic genetic discrimination against persons with genetics-based disabilities.[226]

---

[225] Section 3(1) of Canada's *Genetic Non-Discrimination Act* prohibits the requirement for "an individual to undergo a genetic test., and Section 4(1) prohibits the requirement for "an individual to disclose the results of a genetic test" (Government of Canada, 2017).

[226] Failure of legal recognition of these discriminations against persons with genetics-based disabilities reflects in lack of legal recognition of the pervasive

Regarding employment discrimination, prior to the *Act*, employers could insist that complete family histories of hereditary conditions be included on all applications for long-term employment. Employers justified this insistence, particularly regarding adult-onset genetics-related conditions such as early onset breast cancer,[227] because some employers felt it their right to know this information about potential employees before time and money was invested in their training.[228] Fear of genetic discrimination caused many persons applying for employment to choose not to include this part of their family histories, knowing they needed to conceal adult-onset genetic conditions in order to secure long-term employment.

For life insurance, however, the concealing of hereditary information on application forms is more problematic, as withholding such information voids payment of the policy if the policy holder dies from the concealed condition. The imperative of life insurance companies is to make money for their shareholders and executives, rather than provide financial support for the relatives of deceased policy holders as television advertisements would have us believe. Indeed, prior to Canada's *Genetic Non-Discrimination Act*,[229] most Canadian women at hereditary high risk of breast/ovarian cancer avoided having genetic testing in fear of learning they possessed a BRCA gene mutation and might have to declare such on their life insurance applications.

The extensive knowledge of insurance-company executives regarding the genetics of hereditary conditions, particularly BRCA

---

discrimination against all persons with disabilities; Mykitiuk & Nisker, 2010; Nisker, 2001b, 2012.

[227] See Chapter 6, Miriam; See Chapter 10, She Lived with the Knowledge.

[228] This requirement reminisces the film *Philadelphia* (Demme, 1993), in which the employers of Andrew Beckett, portrayed by Tom Hanks in his Academy Award–winning role, felt they should have been made aware that he was carrying the human immunodeficiency virus (HIV).

[229] Government of Canada, 2017.

gene mutations,[230] was brought to my attention by questions in the "open discussion" following my invited presentation at an international life insurance conference held in Canada.[231]

During the subsequent coffee break, I caffeinated-up the courage to ask one of the conference organizers what he thought about the question regarding mandatory genetic screening. He responded that genetic screening is actually unnecessary because the questions on life insurance application forms regarding family history are just as good as and much less expensive than genetic screens. He added that of course the questions on life insurance applications must be answered honestly or the policy won't be paid out. I asked him further questions, and was blown away by his knowledge of genetics to the point that I asked him if he was a physician or a genetic-scientist. He responded, "Neither, though genetics is very important to life insurance so I must keep up on it, particularly regarding common genetic conditions like BRCA gene mutation breast cancer."[232]

---

[230] See Chapter 7, Miriam; See Chapter 10, She Lived with the Knowledge; See Chapter 15, The Injustice of Needing Angelina Jolie; See Chapter 25, The Arrogance of "But All You Need Is a Good Index Finger"; BRCA gene mutations are autosomal dominant and of high penetrance; "autosomal dominant" means you only need to inherit the gene from one parent to potentially develop the related condition; "high penetrance" means that if you inherit the gene, you are very likely to develop the condition.

[231] One of the questions concerned the ethics of insisting that all applicants for life insurance allow their blood to be screened for genetic conditions in addition to the standard blood glucose determination. I responded that insisting on genetic screening would be coercive, and that refusing life insurance on the basis of such screening was the exact type of discrimination that is prohibited in most countries, and would soon be prohibited in Canada.

[232] See Chapter 5, Princess Margaret; See Chapter 6, Miriam; See Chapter 10, She Lived with the Knowledge; See Chapter 15, The Injustice of Needing Angelina Jolie; See Chapter 18, Victor; See Chapter 25, The Arrogance of "But All You Need Is a Good Index Finger"; Narod, Feunteun, Lynch, Watson, Conway, *et al.*, 1991; Narod, Lynch, Conway, Watson, Feunteun, *et al.*, 1993.

When he mentioned BRCA gene mutations,[233] my discomfort responded by provoking him with, "Are you suggesting women with strong family histories of BRCA-gene mutations related breast cancer should be denied life insurance unless they have prophylactic mastectomies?"[234] He was taken aback, but then smiled and said, "No one ever asked me that before." He paused before continuing, "Life insurance could of course be provided to women with strong family histories of early onset breast cancer without BRCA screening, and without prophylactic mastectomies, but at a much higher rate."[235] "Much higher rate?" I couldn't help but step up on a social-justice soapbox, like my Grandmother used to do,[236] and retort, "Then only women with significant financial means who have a family history of hereditary breast cancer will be able to purchase life insurance." He shrugged. I walked away.

In 2017, as per ritual on my returns to my alma mater, I don my running gear in the Hart House Men's Change Room, where, as a medical student, I had donned soccer cleats, football cleats, and high-cut black Converse All Star basketball shoes,[237] before heading down the hollow-and-high-arched ceilinged hallowed halls. I exit a mock-Gothic door and set off for a slower-than-previous run through campus and beyond. When I was a medical student, these runs were more fun, and I could run faster; however, I still pursue the same route, which loops back to the "Helix of Life," in which all life begins.

---

[233] See Chapter 5, Princess Margaret; See Chapter 6, Miriam; See Chapter 10, She Lived with the Knowledge; See Chapter 15, The Injustice of Needing Angelina Jolie; See Chapter 18, Victor; See Chapter 25, The Arrogance of "But All You Need Is a Good Index Finger."
[234] Nisker, 2007, 2012, 2013; Nisker, Martin, Bluhm, & Daar, 2006.
[235] Nisker, 2007, 2013; Nisker, Martin, Bluhm, & Daar, 2006.
[236] As described later in this chapter, as well as in Chapter 2, "You Must Go to Medical School or Hitler Will Have Won"; also see Chapter 5, Princess Margaret; and Chapter 10, She Lived with the Knowledge.
[237] I commonly see "Converse All Stars" on the everyday feet of university students, men and women alike, in both high-cut and low-cut versions, of black or white, and in a spectrum of fluorescent colours.

I commence running, or more precisely jogging, toward University College, then cut across the flag-football field where our Meds team twice almost won the interfac championship. I cross Harbord Street to dodge through the black-robed students of Trinity College on my way to St. Hilda's, once its women's residence, the walls of which I scaled frequently to my Juliet's window.[238] My lungs fill with fondness and then exhale loss looking at that window, before I continue north to Varsity Stadium, where I run around its track four times, remembering finishing almost last every year. I then head back past the recently ordained Munk School of Global Affairs[239] and work my way south and west to New College, where I reminisce an unethical psych experiment in which all medical students were forced to participate.

Medical school was six years for me, as I was accepted out of high school. In first year, we all had to take Psych 100, which included "volunteering" for a psych experiment.[240] I reported as directed to an old house on St. George Street just south of New College. The house was locked. I knocked but there was no response. I naturally assumed I had reported at the wrong time, and resolved to never come back, when a diminutive, beautiful woman about my age, whom I had gazed longingly at for weeks in the New College Library, came up the steps. She softly asked if the door was locked. I had no words to respond, as I was too busy thanking the spiritverse for the good fortune of this coincidence.

Too soon a psych grad student opened the door and took us to separate rooms. I was handed a sheet of paper instructing me how to behave during the experiment. I was to be a typical jock, which bothered me, as my personality should not have been transparent to the grad student, and use my excess strength to resist the push of my

---

[238] Shakespeare, 1734.

[239] The Munk Centre for International Studies was founded in 2000 and was renamed the Munk School of Global Affairs in 2010. In 2018, it merged with University of Toronto's School of Public Policy and Governance to become the Munk School of Global Affairs and Public Policy (personal communication, July 22, 2021).

[240] See Appendix I, They Psych Experiment

partner. I again thanked the spiritverse for the upcoming touch of this woman, just as a door opened into the one-way-mirrored observation room in which our encounter would occur. My partner appeared shy, and my pounding heart forced my eyes to look down at the floor rather than into her eyes. I heard the grad student's intercom calm voice say "Start," and my partner pushed the entirety of her small body into my chest, soon getting out of her sweet breath. The grad student then unfortunately came in and then led us back to our previous rooms to fill out a lengthy multiple-choice questionnaire. He told us to leave whenever we were finished. Thanking the spiritverse again, I quickly ticked the boxes without reading the questions and ran out the front door to wait for the woman I would be asking for a date for that very night. I waited, I waited.[241]

I continue south on Spadina Avenue, circling the decaying architecture of Connaught Laboratories where, in its glory days, insulin was invented[242] and the Salk vaccine manufactured.[243] Halfway around the circle's perimeter, I stop reverently at the funeral parlour

---

[241] I assumed that my partner's questionnaire was either much longer than mine or that she took the questions more seriously than I did. After a half hour of waiting, I began thinking there might be another door, and ran around to the back door, but there was no one there. I ran back to the front door, only to find a classmate of mine waiting to go in to start his experiment. There didn't seem to be a partner there for him. Still thanking the spiritverse, I looked forward to Monday morning in New College Library where I planned to stay all day until my date for next weekend, and maybe forever, came in, but she didn't. On Tuesday, the classmate I saw waiting to go in for his psych experiment, came into the library and I told him I was waiting for the love of my life to come and sit where she always sat. I started describing her appearance in detail until my classmate stopped me with, "Wait a minute, you mean that real bitch who was my psych-experiment partner?"

[242] Frederick Banting, insulin's co-inventor, had been a physician in London, Ontario, before moving to the University of Toronto. Banting's co-inventor was Charles Best, a medical student working with Banting as a summer job (Bliss, 1982).

[243] See Chapter 19, Canadian COVID Injustice on Beaches and Beach Volleyball Courts

from where I helped carry my Grandmother's coffin through a sea of mourners.[244] I then accelerate east on College Street past the Clarke Institute, a tallish white building named for Canada's infamous eugenicist Charles Clarke, who campaigned for a pure Canada through sterilization of "the defective"[245] and supported Alberta's *Sexual Sterilization Act*.[246] Clarke also campaigned for a pure Canada by arguing that Canada should refuse to accept Jews who had survived Hitler's concentration camps. The Clarke Institute's name was changed in 1998 after reports on Clarke's eugenics campaigns persisted in the press.[247]

In the 1990s, another Canadian eugenicist and pseudoscientist, Philippe Rushton,[248] a psychology professor at Western University where I work, went so far as to claim that Black people performed poorer on intellect tests because they had smaller brains, as he tried to demonstrate by measuring their supposedly smaller skulls.[249] In response to Rushton's "scientific" racism,[250] students of African

---

[244] See Chapter 2, "You Must Go to Medical School or Hitler Will Have Won"; See Chapter 5, Princess Margaret; See Chapter 10, She Lived with the Knowledge.

[245] Dowbiggin, 2005.

[246] Alberta's *Sexual Sterilization Act* was repealed in 1972, when I was in medical school (Marshall & Robertson, 2006).

[247] Dr. Clarke was the University of Toronto's first Professor of Psychiatry (1908–1924) and was the first Director of the Toronto Psychiatric Hospital, which, in 1966, was replaced by the Clarke Institute of Psychiatry, named for him (Temerty Faculty of Medicine, 2021). The Clarke Institute is now the Centre for Addiction and Mental Health (CAMH) (Dowbiggin, 2005). When thinking about the expunging of Clarke from the University of Toronto, I can't help but feel the resonation of the pulling down of statues of John A. Macdonald, Canada's first prime minister because of his cruel policies against our First Nations and Metis peoples (Hristova, 2021).

[248] Rushton, 1995; Western University, 2020.

[249] Rushton, 1995.

[250] Scientific racism has existed for centuries, both in the unfortunate research of well-intentioned anthropologists such as Margaret Mead (1972) and Bronislaw Malinowski (1922), and in the fortune-making Inuit exploitations (Rivet, 2019) of entrepreneurs such as Carl Hagenbeck.

descent at Western started wearing tape-measure sweat bands around their heads.[251]

I turn left into an alley and run past Engineering and Convocation Hall, then do two ritual laps around the soccer field, where our black-and-red–striped "Meds" team lost more games than we won. I then duck through the tunnel under Queen's Park Circle and do two ritual laps on the trail around the Provincial Legislature's massive pink stones, before proceeding to the Victorian mansions on the eastern section of campus, in which once-elegant rooms have been converted to classrooms, common rooms, study rooms, and profs' offices for Victoria College and St. Michael's College.

I then leave campus continuing east past more Victorian mansions, in which other once-elegant rooms had been subdivided to multiple rooming-house rooms for Toronto's impoverished, but recently have been reconverted to gentrified rooms in trendy townhouses for Toronto's "upwardly mobile."[252] I occasionally make short diversions into small side streets, where tiny plaque-posted "century cottages" proclaim the past of Toronto's workers who laboured for the owners of the mansions. Soon there will be neither rooms nor room for the socio-economically disadvantaged here, not even in the tall social-assistance apartments of St. James Town,[253] nor in the back alleys in its shadow.[254]

At Parliament Street,[255] named thus because Canada's Parliament once graced this now impoverished street, I turn north across Wellesley Street, before jutting back a block to Sherbourne, where

---

[251] Mahood, 1989; Woods, 1989. Our University could not fire Rushton because he had tenure, and his presence remained an embarrassment to all of us. Although he stopped teaching in the early 1990s, he continued to conduct "research" as a tenured professor until his death in 2012 (Western University, 2020).

[252] See Chapter 14, Beneath the BMW's Wheels.

[253] St. James Town West Park is to be demolished in 2022 (City of Toronto, 2021).

[254] See Chapter 14, Beneath the BMW's Wheels.

[255] See Chapter 14, Beneath the BMW's Wheels.

Princess Margaret Hospital stood when I was a medical student.²⁵⁶ Princess Margaret was where I rotated to learn cancer care,²⁵⁷ but, emotionally more important, was where my Grandmother endured treatments for her early onset breast cancer six years before.²⁵⁸ As a medical student, I always felt deflated as I stopped solemnly at the steps up to Princess Margaret's entrance. The visiting-prof me still feels deflated as I stop where Princess Margaret's steps once stood.

I pump myself up to proceed east to the Bloor Street Viaduct.²⁵⁹ However, I do not cross the bridge, because when I was in high school, a friend jumped from it before the safety screens were erected due to the bridge becoming such a common site for suicides.²⁶⁰ Instead of crossing the viaduct, I go down the ramp heading south on the Don River Trail. As a medical student I never turned around until the infamous Don Jail, but the visiting-prof me turns around long before, and has to work hard to lumber back to campus. When at last back on campus, I head to its most important edifice, and run up the Medical School's steps, fighting gasps of collapse like Rocky running up the steps of the Philadelphia Museum of Art when he first started

---

[256] It was originally named the Ontario Cancer Institute in 1952. In 1958, it was renamed Princess Margaret Hospital and was officially opened by Her Royal Highness the Princess Margaret. In 1996, Her Royal Highness the Princess Margaret officially re-opened it at its current location on University Avenue. In 1998, Princess Margaret Hospital became part of the University Health Network, and in 2012 it was renamed Princess Margaret Cancer Centre (University Health Network, 2021); See Chapter 5, Princess Margaret.

[257] See Chapter 5, Princess Margaret.

[258] See Chapter 5, Princess Margaret; See Chapter 10, She Lived with the Knowledge.

[259] The bridge was originally erected as a "viaduct" to transport water pipes across Toronto. Michael Ondaatje describes the building of the bridge in *In the Skin of a Lion* (Ondaatje, 1987). The Bloor Street Viaduct is also featured in the Barenaked Ladies' song "War on Drugs" (Barenaked Ladies, 2003) and Bruce Cockburn's song "Anything Could Happen" (Cockburn, 1988).

[260] See Chapter 5, Princess Margaret.

training.²⁶¹ Still gasping, I raise my arms in Victory's V, like Rocky did when he had been training for a while, but I do this only briefly, as I am confronted with the "Helix of Life."

The Helix was the site of my first medical-student, social-justice protest, a protest that had little to do with the interrogation of the discrimination that unbridled new genetic technologies would foster against persons with genetics-related disabilities that would become my research focus, and everything to do with the press's interrogation of our then premier's accused corruption. The premier had just awarded a lucrative construction contract without tender to one of his recent election campaign's largest donors. The new Ontario Power Building's "gleaming" glass²⁶² would soon be rising to stand kitty-corner to both the Provincial Legislature's hundred-year-old building and our Medical School's brand-new building. The Medical School's new building was constructed for class sizes twice as large as mine, to contribute to the predicted vast increase in the number of physicians that would be required to care for the imminent explosion of our aging population. Although the explosion in our aging population did occur, the class sizes were cut in half by the premier's tax-break–elected government's funding cutbacks that, along with other provincial Premier's cutbacks, left Canada with one of the lowest physician per capita ratios of "developed" countries.²⁶³

The morning of our social-justice protest, I learned that the edifice in which we were learning medicine would be officially opened at two o'clock by that less-than-social-justice premier. This idealistic medical student could not resist an attempt to inhibit the official opening of our high-purpose edifice by a premier under snow clouds of corruption.²⁶⁴ I stared down from the windows of the

---

²⁶¹ Rocky was portrayed by Sylvester Stallone in director John Avildsen's 1976 Academy Award–winning film (Avildsen, 1976).
²⁶² Paikin, 2016, p. 172.
²⁶³ Nisker, 2018, 2019; World Bank Group, 2018.
²⁶⁴ This premier had a social conscience when compared with another premier of his political stripe, who gutted our hospital and universities, and

corridor outside the physiology labs onto the "Helix of Life" as I considered how to inhibit the premier's imminent official opening. The Helix was hard to see that day, as it had been covered with a semi-transparent plastic sheath, tied tightly in place with an enormous red ribbon. I assumed the sheath stretched over the Helix was meant to keep the snow off, but to this not-far-from-adolescence medical student, the sheath looked like a giant condom stretched over a hugely erect penis. Suddenly, I had the "how" of my social-justice protest and ran down the steps to the first-year lecture theatre that I usually avoided, to catch my class before it was dismissed for lunch. I needed volunteers to help me roll two gigantic snowballs and station them beside the "Helix of Life" in the "anatomically correct" position.[265]

---

who was responsible for the death of Dudley George at the Kettle and Stony Point First Nation protest at Ipperwash Beach, 45 minutes from my home (Dubinski, 2020). Following Dudley George's assassination, I was recruited by Rev. Susan Eagle to be her "White doctor spokesperson" for her other social-justice protests. Rev. Eagle took great pride in squishing a cream pie into the premier's face in front of the Ontario Legislative Building. Rev. Eagle became famous when standing behind the Oka barricade in Quebec (Hamilton, 1990).

[265] As soon as the prof exited the always much-less-than-half-full lecture theatre, I dashed to the front of the still-yawning class and asked for volunteers to stop the premier from officially opening our school. Cheers of "Yes" accompanied classmates bounding down the lecture theatre's steps. Approximately 50 of us put on our jackets and marched with purpose across the soccer field to the side distant from the "Helix." There I made two palm-size snowballs and handed them to classmates. These eager snowman-builders proceeded to roll the accelerating-in-size snowballs across the soccer field until their snowballs became too heavy for one person to push. They enlisted other classmates, who eventually sat on shoulders to push forward our enormous snowballs. When the snowballs abutted the steps up to the "Helix," it took almost all of us, pushing one snowball up at a time, to struggle them up the steps. Next, it was necessary to station each snowball in the "anatomically correct" position. I dashed through our Medical School's main entrance, ran up the steps to the physiology corridor's windows, and with arm motions directed the snowballs' positioning.

Our after-lunch lecture was presented to even fewer students than usual, as 50 of us were looking down from the physiology corridor's windows at our enhanced Helix. Soon four men in black trench coats appeared and started circling the sheathed Helix, scratching their heads, staring quizzically and talking rapidly to each other. Suddenly, one of the men grabbed his head aghast, and spoke urgently into his hand. A few minutes later, a black-tarpaulined U of T maintenance truck pulled up, and eight men with shovels disembarked from its back. The men went hard at our snowballs, urged on to go faster by the black-coated representatives of our premier. The shovellers did their best, but made little progress. When two o'clock came and went, we cheered, high-fived, jumped around, hugged each other. About half an hour later a large red tractor chugged up, spewing black fumes. The men with shovels dropped their useless tools and yanked enormous chains from a large box attached to the back of the tractor. The larger left snowball was chained to the tractor first, and dragged to the centre of the soccer field that still exhibited the tracks engraved by our snowballs. The tractor returned and repeated its painful pull. Then the men with shovels transformed into men with brooms, and they vigorously swept the steps and other surfaces surrounding the Helix. When the surfaces were pristine, a black limousine pulled up, and one of the black-coats spoke vigorously into his hand. A few seconds later another black-coat opened the limo door, and the premier surfaced to boos he could not hear because the physiology corridor's windows could not be opened without breaking their glass; a suggestion on which we decided to pass. The premier smiled broadly as he waved at cameras, then posed beside the Helix with cartoon-like large blue scissors. The premier quickly cut the ribbon, and the Helix was unsheathed by his security men. The premier's picture was quickly taken with the naked Helix before he was just as quickly ushered back into his limo to go back under Queen's Park Circle to his office.

The unsheathed Helix became increasingly less humorous over the years of the unsheathing of unregulated reproductive-genetics research[266] and "therapies,"[267] and my growing appreciation of this danger, including in my six-year role as Co-Chair of Health Canada's Advisory Committee on New Reproductive Genetic Technologies that informed Canada's *Assisted Human Reproduction Act*.[268] I found no humour as I stood transfixed before the Helix's DNA spiral 46 years after its planting, as I was consumed by the flame of the discrimination that the accelerating "bright future" of DNA research and treatment[269] fosters against persons living with genetics-related disabilities; persons whose disabilities could have been prevented by preventing their births.[270] As I stood under the Helix in 2017, I contemplated how Canada's new *Genetic Non-Discrimination Act*[271] could have missed the opportunity to legislate "genetic non-discrimination" against persons with genetics-related disabilities beyond employment and insurance discrimination.

I must always labour up the steps to the Helix, as I was a culprit in the early days of reproductive genetics, researching preimplantation genetic diagnosis (PGD),[272] albeit for the honourable purpose of offering an alternative to amniocentesis for women already undergoing the rigours of IVF.[273] However, in 1991, someone leaked to the press what we were researching, triggering calls from across Canada requesting PGD; 57 percent were for the purpose of ensuring a male baby. I closed my laboratory, returned the research grant money, and devoted my future research to the ethics and health-policy issues emanating from PGD.[274] However, new applications of PGD exploded

---

[266] Nisker, 2001b, 2012, 2021; Nisker & Gore-Langton, 1995.
[267] Baylis, 2013; Lane & Nisker, 2016; Nisker, 2015, 2021.
[268] Government of Canada, 2004.
[269] Treatment is still, in most cases, the precaution of the birth of persons with disabilities.
[270] Nisker, 2001a, 2012, 2018; See Chapter 17, Ruth.
[271] Government of Canada, 2017.
[272] Handyside, Pattinson, Penketh, Delhanty, Winston, *et al.*, 1989; Nisker, 2012; Nisker & Gore-Langton, 1995.
[273] Nisker, 2001b, 2012; Nisker & Gore-Langton, 1995.
[274] Lane & Nisker, 2016; Mykitiuk & Nisker, 2010; Nisker 2012, 2015, 2021; Nisker & Gore-Langton, 1995.

faster than our research could address, including for "therapeutic" purposes, and, even more worrisome, for enhancement purposes.[275]

Misnomers[276] began camouflaging reproductive-genetics research and clinical procedures from the eyes of national regulators, ethics committees, and the general public; research and procedures that seemed science fiction, but in reality were occurring in Canada, albeit illegally.[277] "Mitochondrial replacement therapy,"[278] an example of such a misnomer, has nothing to do with "replacement" of mitochondria in the oocytes of women carrying mitochondrial genes that could be phenotypically expressed in their sons. Rather, "mitochondrial replacement therapy" is a misnomer employed to camouflage germ-line nuclear transfer, a reproductive-cloning technology illegal in countries where legislation regulates reproductive genetics technologies.[279]

---

[275] Baylis, 2013; Lane & Nisker, 2016; Mykitiuk & Nisker, 2010; Nisker, 2015.
[276] See Chapter 16, A Brief and Personal History of "What's in a Name."
[277] Baylis, 2013, 2019; Lane & Nisker, 2016; Nisker, 2015, 2021.
[278] For mitochondrial replacement therapy (MRT), oocytes of a socio-economically disadvantaged woman, frequently in Eastern Europe or an impoverished community in the United States, are surgically removed through IVF needles after "hyperstimulation" of their ovaries with menotropin drugs (Abramov, Elchalal, & Schenker, 1999; Balen & Wing, 2005; Nisker, 1997, 2001b; Practice Committee of the American Society for Reproductive Medicine, 2003). The nucleus of the purchasing (usually wealthy) woman's oocyte, also obtained through IVF technology, is then inserted into the now enucleated cytoplasm of the oocyte of the "donor," another misnomer. This insertion is followed by the spermatozoa insertion through intracytoplasmic sperm injection (ICSI) (Van Steirteghem, Nagy, Liu, Joris, Verheyen, *et al.*, 1994) of the wealthy woman's partner or a paid sperm "donor" into the "donor" woman's oocyte. The purpose of this risk of harm to a socio-economically disadvantaged woman is so the wealthy woman/couple can avoid adoption (Nisker, 2001b). A physician's participation in MRT breaks the Hippocratic Oath of *primum non nocere*, first do no harm, as harm is being done to the "donor" woman, even despite her receiving financial compensation (Nisker, 1997, 2001b).
[279] Baylis, 2013; Lane & Nisker, 2016; Mykitiuk & Nisker, 2010; Nisker, 2015.

Spiralling-out-of-control applications of DNA screening are fostered by increasingly less expensive and less-risk-of-harm methods of DNA screening, such "non-invasive prenatal testing" (NIPT).[280] For NIPT, the DNA of fetal cells circulating in the blood of a pregnant woman are centrifuged down and subjected to microarray analysis. If the analysis of maternal blood cells indicates the fetus has Down syndrome or another "anomaly," the fetus is usually aborted.[281] The indications for NIPT have expanded beyond the original criteria of a strategy to be offered as a step between a concerning integrated pregnancy screen (IPS) and amniocentesis to now be considered "routine" in all pregnancies.[282] However, there has not been extensive ethics research regarding the effect of routine NIPT on persons living with genetics-related disabilities, nor is there likely to be in the future, as this barn door to embryo selection has been pried wide open, and will be difficult to close.

The cement of "Helix of Life" has been further softened recently by CRISPR (clustered regularly interspaced short palindromic repeats)[283] and other PGD-related "gene-editing" technologies that scientists are rapidly developing to modify embryos.[284] I am concerned that these technologies, and other front-page genetic strategies proclaimed to improve the human race,[285] will promote the imperative for wealthy

---

[280] See Chapter 16, A Brief and Personal History of "What's in a Name"; Langlois, Brock, & Genetics Committee, 2013; Murdoch, Ravitsky, Ogbogu, Ali-Khan, Bertier, *et al.*, 2017; Prenatal Screening Ontario, 2021; Nshimyumukiza, Beaumont, Duplantie, Langlois, Little, *et al.*, 2018; Vanstone, King, de Vrijer, & Nisker, 2014; Vanstone, Kinsella, & Nisker, 2012; Vanstone, Yacoub, Winsor, Giacomini, & Nisker, 2015; Zwingerman & Langlois, 2020.
[281] Vanstone, King, de Vrijer, & Nisker, 2014; Vanstone, Yacoub, Winsor, Giacomini, & Nisker, 2015.
[282] Langlois, Brock, & Genetics Committee, 2013; Vanstone, Yacoub, Winsor, Giacomini, & Nisker, 2015.
[283] Ran, Hsu, Wright, Agarwala, Scott, *et al.*, 2013.
[284] Baylis, 2019.
[285] White, 2016.

persons to modify their embryos in order to pursue what our blinkered-society privileges as ideal characteristics.[286]

Improvement in the human condition does not require genetic manipulation; rather improvement in social determinants of health,[287] including education, nutrition, housing, social inclusion, laws, and healthcare,[288] as well as a change in the way we see and accommodate persons with disabilities.[289] We must work for further anti-discrimination legislation that will address the context and purpose of genetic screening of embryos and fetuses, as the DNA variations and phenotypic differences that we all possess contribute to the diversity that makes our society richer.[290] We must embrace each other as being "more than the sum of our scripted genes,"[291] and embrace the concept that all of us are less when our worth is determined by our genetics.

A new generation of socially conscious medical students and clinicians must engage in social-justice protests to make Canada more tolerant, more accommodating, and more respectful of difference.

A portion of an early short version of this chapter was published as "The Cement Spiral" in the Journal of Obstetrics and Gynaecology Canada in June 2018.[292]

---

[286] See Chapter 1, The Rotor; See Chapter 3, I'm Sorry Ronnie; Mykitiuk & Nisker, 2010; Nisker, 2001b, 2012.
[287] Mykitiuk & Nisker, 2010.
[288] Frazee, Gilmour, & Mykitiuk, 2006; Frazee, Gilmour, Mykitiuk, & Bach, 2002; Mykitiuk & Nisker, 2010; Raphael, 2004; Siegler & Epstein, 2003.
[289] World Health Organization, 1948.
[290] Nisker, 2001b, 2012.
[291] Nisker, 2001b, 2012.
[292] Nisker, 2018.

References

Abramov, Y., Elchalal, U., & Schenker, J. G. (1999). Severe OHSS: An "epidemic" of severe OHSS: A price we have to pay? *Human Reproduction, 14*(9), 2181–2183.

Avildsen, J. (Director). 1976. *Rocky*. United Artists.

Balen, A., & Wing, C. (2005). *Ovarian hyperstimulation syndrome—A short report for the HFEA*. Human Fertilisation and Embryology Authority. http://data.parliament.uk/DepositedPapers/Files/DEP2008-2052/DEP2008-2052.pdf

Barenaked Ladies. (2003). War on drugs. On *Everything to everyone*. Reprise.

Baylis, F. (2013). The ethics of creating children with three genetic parents. *Reproductive BioMedicine Online, 26*(6), 531–534.

Baylis, F. (2019). *Altered inheritance: CRISPR and the ethics of human genome editing*. Harvard University Press.

Bliss, M. (1982). *The discovery of insulin*. McClelland and Stewart.

City of Toronto. (2021). St. James Town West Park redesign. https://www.toronto.ca/city-government/planning-development/construction-new-facilities/improvements-expansion-redevelopment/st-james-town-west-park-revitalization/

Cockburn, B. (1988). Anything can happen. On *Big circumstance*. True North Records.

Demme, J. (Director) (1993). *Philadelphia*. TriStar Pictures.

Dowbiggin, I. (2005). Clarke, Charles Kirk. In *Dictionary of Canadian biography*, vol. 15. University of Toronto/Université Laval. http://www.biographi.ca/en/bio/clarke_charles_kirk_15E.html

Dubinski, K. (2020, Sept 6). 25 years after his death, Dudley George's fight for the land continues. Canadian Broadcasting Corporation News. https://www.cbc.ca/news/canada/london/dudley-george-kettle-stony-point-25-anniversary-1.5708055

Frazee, C., Gilmour, J., & Mykitiuk, R. (2006). Now you see her, now you don't: How law shapes disabled women's experience of exposure, surveillance, and assessment in the clinical encounter. In D. Pothier & R. Devlin (Eds.), *Critical disability theory: Essays in philosophy, politics, policy and law* (pp. 223–247). University of British Columbia Press.

Frazee, C., Gilmour, J., Mykitiuk, R., & Bach, M. (2002). The legal regulation and construction of the gendered body and of disability in Canadian health law and policy. *National Network on Environments and Women's Health.* http://www.cwhn.ca/en/node/24745

Government of Canada. (2004). *Assisted Human Reproduction Act* (S.C. 2004, c. 2). https://laws-lois.justice.gc.ca/eng/acts/A-13.4/page-1.html#h-6052

Government of Canada. (2017). *Genetic Non-Discrimination Act* (S.C. 2017, c. 3). https://laws-lois.justice.gc.ca/eng/annualstatutes/2017_3/

Hamilton, J. (1990, Sept 10). Church ministers support Mohawks. *London Free Press*, A3.

Handyside, A. H., Pattinson, J. K., Penketh, R. J., Delhanty, J. D., Winston, R. M., & Tuddenham, E. G. (1989). Biopsy of human preimplantation embryos and sexing by DNA amplification. *Lancet, 1*(8634), 347–349.

Hristova, B. (2021, Aug 16). Removing John A. Macdonald statue "tip of the iceberg" for reconciliation: Hamilton organizer. Canadian Broadcasting Corporation News. https://www.cbc.ca/news/canada/hamilton/statue-1.6142346

Lane, A., & Nisker, J. (2016). "Mitochondrial replacement" technologies and human germline nuclear ,odification. *Journal of Obstetrics and Gynaecology Canada, 38*(8), 731–736.

Langlois, S., Brock, J. A., & Genetics Committee. (2013). Current status in non-invasive prenatal detection of Down syndrome, trisomy 18, and trisomy 13 using cell-free DNA in maternal plasma. *Journal of Obstetrics and Gynaecology Canada, 35*(2), 177–181.

Mahood, C. (1989, February 14). Rushton theories clash with provincial race policies. *The Gazette*, 1.

Malinowski, B. (1922). *Argonauts of the Western Pacific*. E. P. Dutton & Co.

Marshall, T., & Robertson, G. (2006). Eugenics in Canada. In *The Canadian Encyclopedia*. Historica Canada. https://www.thecanadianencyclopedia.ca/en/article/eugenics

Mead, M. (1972). *Blackberry winter: My earlier years*. HarperCollins Publishers.

Murdoch, B., Ravitsky, V., Ogbogu, U., Ali-Khan, S., Bertier, G., Birko, S., Bubela, T., De Beer, J., Dupras, C., Ellis, M., Granados Moreno, P., Joly, Y., Kamenova, K., Master, Z., Marcon, A., Paulden, M., Rousseau, F., & Caulfield, T. (2017). Non-invasive prenatal testing and the unveiling of an impaired translation process. *Journal of Obstetrics and Gynaecology Canada*, *39*(1), 10–17.

Mykitiuk, R., & Nisker, J. (2010). Social determinants of "health" of embryos. In J. Nisker, F. Baylis, I. Karpin, C. McLeod, & R. Mykitiuk (Eds.), *The "healthy" embryo: Social, biomedical, legal and philosophical perspectives* (pp. 116–135). Cambridge University Press.

Narod, S., Lynch, H., Conway, T., Watson, P., Feunteun, J., & Lenoir, G. (1993). Increasing incidence of breast cancer in family with BRCA1 mutation. *Lancet*, *341*(8852), 1101–1102.

Narod, S. A., Feunteun, J., Lynch, H. T., Watson, P., Conway, T., Lynch, J., & Lenoir, G. M. (1991). Familial breast-ovarian cancer locus on chromosome 17q12-q23. *Lancet*, *338*(8759), 82–83.

Nisker, J. (2001). Orchids: Not necessarily a gospel. In J. Murray (Ed.), *Mappa mundi: Mapping culture/mapping the world* (pp. 61–110). University of Windsor Press. http://www.uwindsor.ca/hrg/mappa-mundi-mapping-culturemapping-the-world-table-of-contents-0

Nisker, J. (2012). *From Calcedonies to Orchids: Plays promoting humanity in health policy*. Iguana Books.

Nisker, J. (2013). A public health education initiative for women with a family history of breast/ovarian cancer: Why did it take Angelina Jolie? *Journal of Obstetrics and Gynaecology Canada, 35*(8), 689–691.

Nisker, J. (2015). The latest thorn by any other name: Germ-line nuclear transfer in the name of "mitochondrial replacement." *Journal of Obstetrics and Gynaecology Canada, 37*(9), 829–831.

Nisker, J. (2018). The cement spiral. *Journal of Obstetrics and Gynaecology Canada, 40*(6), 643–645.

Nisker, J. (2019). Dissolution of Canada's single-tiered health system would threaten the health of women with disabilities. *Journal of Obstetrics and Gynaecology Canada, 41*(11), 1616–1618.

Nisker, J. (2021). A brief and personal history of "what's in a name" in reproductive genetics. *Medical Humanities, 47*(2), 228–234.

Nisker, J., Martin, D. K., Bluhm, R., & Daar, A. S. (2006). Theatre as a public engagement tool for health-policy development. *Health Policy, 78*(2–3), 258–271.

Nisker, J. A. (1997). In quest of the perfect analogy for using in vitro fertilization patients as oocyte donors. *Womens Health Issues, 7*(4), 241–247.

Nisker, J. A. (2001a). Chalcedonies. *Canadian Medical Association Journal, 164*(1), 74–75.

Nisker, J. A. (2001b). Physician obligation in oocyte procurement. *American Journal of Bioethics, 1*(4), 22–23.

Nisker, J. A. (2007). The need for public education: "Surveillance and risk reduction strategies" for women at risk for carrying BRCA gene mutations. *Journal of Obstetrics and Gynaecology Canada, 29*(6), 510–511.

Nisker, J. A., & Gore-Langton, R. E. (1995). Pre-implantation genetic diagnosis: A model of progress and concern. *Journal of Obstetrics and Gynaecology Canada, 17*(3), 247–262.

Prenatal Screeing Ontario. (2021). Non-invasive prenatal testing. Better Outcomes Registry and Network (BORN) Ontario.

https://prenatalscreeningontario.ca/en/pso/about-prenatal-screening/non-invasive-prenatal-testing.aspx

Nshimyumukiza, L., Beaumont, J. A., Duplantie, J., Langlois, S., Little, J., Audibert, F., McCabe, C., Gekas, J., Giguère, Y., Gagné, C., Reinharz, D., & Rousseau, F. (2018). Cell-free DNA-based non-invasive prenatal screening for common aneuploidies in a Canadian province: A cost-effectiveness analysis. *Journal of Obstetrics and Gynaecology Canada*, 40(1), 48–60.

Ondaatje, M. (1987). *In the skin of a lion*. McClelland and Stewart.

Paikin, S. (2016). *Bill Davis: Nation builder, and not so bland after all*. Dundurn.

Practice Committee of the American Society for Reproductive Medicine. (2003). Ovarian hyperstimulation syndrome. *Fertility and Sterility*, 80(5), 1309–1314.

Ran, F. A., Hsu, P. D., Wright, J., Agarwala, V., Scott, D. A., & Zhang, F. (2013). Genome engineering using the CRISPR-Cas9 system. *Nature Protocols*, 8(11), 2281–2308.

Raphael, D. (2004). Introduction to the social determinants of health. In D. Raphael (Ed.), *Social determinants of health: Canadian perspectives* (pp. 1–18). Canadian Scholars Press.

Rivet, F. (2019). Abraham Ulrikab. In The Canadian Encyclopedia. Historica Canada. https://www.thecanadianencyclopedia.ca/en/article/abraham-ulrikab

Rushton, J. P. (1995). *Race, evolution, and behavior: A life history perspective*. Transaction Books.

Shakespeare, W. (1734). *Romeo and Juliet. By Mr. William Shakespeare*. Eighteenth Century Collections Online. Gale. https://www.gale.com/primary-sources/eighteenth-century-collections-online

Siegler, M., & Epstein, R. A. (2003). Organizers' introduction to the Symposium on Quality Health Care. *Perspectives in Biology and Medicine*, 46(1), 1–4.

Temerty Faculty of Medicine. (2021). Our history. Department of Psychiatry, University of Toronto. https://www.psychiatry.utoronto.ca/our-history

University Health Network. (2021). Princess Margaret history. https://www.uhn.ca/corporate/AboutUHN/OurHistory/Pages/princess_margaret_history.aspx

Van Steirteghem, A., Nagy, Z., Liu, J., Joris, H., Verheyen, G., Smitz, J., Tournaye, H., Liebaers, I., & Devroey, P. (1994). Intracytoplasmic sperm injection. *Baillière's Clinical Obstetrics and Gynaecology, 8*(1), 85–93.

Vanstone, M., King, C., de Vrijer, B., & Nisker, J. (2014). Non-invasive prenatal testing: Ethics and policy considerations. *Journal of Obstetrics and Gynaecology Canada, 36*(6), 515–526.

Vanstone, M., Kinsella, E. A., & Nisker, J. (2012). Information-sharing to promote informed choice in prenatal screening in the spirit of the SOGC clinical practice guideline: a proposal for an alternative model. *Journal of Obstetrics and Gynaecology Canada, 34*(3), 269–275.

Vanstone, M., Yacoub, K., Winsor, S., Giacomini, M., & Nisker, J. (2015). What is "NIPT"? Divergent characterizations of non-invasive prenatal testing strategies. *AJOB Empirical Bioethics, 6*(1), 54–67.

Western University. (2020). Dr. Philippe Rushton, 1943–2012. Faculty remembrance. Department of Psychology. https://psychology.uwo.ca/people/faculty/remembrance/rushton.html

White, C. (2016). *Human* gene editing and eugenics The Centre for Bioethics and Culture Network. http://www.cbc-network.org/2016/03/human-gene-editing-and-eugenics/

Woods, C. (1989, March 10). *Displaying a head for business* [Photograph]. The Gazette, 2.

World Bank Group. (2018). *Physicians (per 1,000 people)*. World Bank data.

https://data.worldbank.org/indicator/SH.MED.PHYS.ZS?end=2015&locations=CA&start=1961&view=chart

World Health Organization. (1948). *Preamble to the Constitution of the World Health Organization as adopted by the International Health Conference.* World Health Organization.

Zwingerman, R., & Langlois, S. (2020). Committee Opinion No. 406: Prenatal testing after IVF with preimplantation genetic testing for aneuploidy. *Journal of Obstetrics and Gynaecology Canada, 42*(11), 1437–1443.

Chapter 14

Beneath the BMW's Wheels[293]

A large hand reaches out to me,
Up to me, beseeching me
In a gentrifying alleyway,
Strewn with needles and luxury vehicles.
The hand is thickened by scabrous lesions,
Inflammation, angry skin;
As blurred words slur "Sorry,"
"Not feeling well," "Help."
I reach for the hand and see a large man
Collapse further under parked wheels,
Before I could catch him,
Before I could ask him his name.

---

[293] *Beneath the Wheel* is a Hermann Hesse (1968) novel exploring his suppression under an archaic Swiss education system in the early 1900s. I borrow from Hesse's title to recount the injustice inflicted on socio-economically suppressed persons in Canada.

I kneel to better reach
The man beneath the wheels,
Take his hand, tell him "I've got you,"
"I'm going to help you," "I'm a doctor."
I place my left hand on his forehead,
Thickened with more torturous skin,
Sandwiching swollen eyes
Above roughened cheeks, remnants of beard,
And the tattered woollen collar
Of his not-warm-enough red plaid jacket,
Missing buttons on his chest,
Still wet from last night's weather.

I encourage him from beneath the wheels
For better air for him to breathe,
And for me to better assess
The texture of his infirmity.
His watery eyes greet mine,
I reassure, ask his name.
Lack-of-breath doesn't answer,
So I suggest if in pain
He nod his head, which he does,
As I explain he needs a hospital.
His troubles' tremble shakes my hand,
As the man nods again.

Just then a passerby
On his way to a Porsche,
Spits, "These guys clutter our lane,
And should be arrested of course."
And when behind his Porsche's
Clear-coated secure door,
Lowers a tinted window with,
"I'll call the police as I've done before,
But the police can't keep them
Off our streets;
Just tell me to ignore them,
But that's not easy, I park here."

I plead with the Porsche man,
"Please call an ambulance instead."
The man beneath the wheels
Squeezes my hand, nods his head.
But we don't trust the Porsche man,
Driving off, spraying slush;
So I let go of my touch, "Just for a minute,"
Find my phone, push 911.
When asked for the address
To which the ambulance should be sent,
I describe "An alley west of Parliament Street,
Just south of the apartments."

I'm put on hold for several minutes,
Then told, "Police were dispatched."
I demand, "This man needs an ambulance,
And I wouldn't have called if he didn't;"
And add, "I'm a doctor,"
As I hear, "A car will be there."
So I cover the man with my winter coat,
And stroke his furrowed forehead.
The anguish of his blistered lips
Quiver, then whisper,
"Thanks," as he shivers,
As he waits, as he waits.

After too-many too-long minutes,
A police car ambles the alley,
The policeman in it screaming,
"Get away from him, are you crazy,
Do you have any idea
What that guy might have?"
I answer, "Yes, I'm a doctor,"
And hear, "Then you should know better."
He throws out the window,
"At least put on these gloves and mask."
I don't take them as the man's hand asks
To keep holding touch.

The policeman calls an ambulance
That bides its time before arriving;
Its mission for this person
Diverted by "indigent," less urgent.
The man is pulled from between the wheels,
Lifted on a stretcher, loaded in.
I ask permission to ride with him
To ensure he's looked after,
Rather than endure the long internment
Sentenced to invisible persons like him;
Persons seen as "indigent," "indigenous," "chronic,"
Triaged from the focus of the too-few nurses.

But I'm denied permission to ride with him
And don't even ask where he'll be taken,
So I can meet the man there,
And ease his despair of waiting,
Curtained in some corner,
Ignored, enduring pain.
But I don't ask because I fear
The mire of the system,
And I'm a visitor here
With research to finish,
And know in this minute
I'm forever diminished by my silence.

I just stand frozen in the alley
Behind Parliament Street stores,
Watching an ambulance slush north
On its course to some Emerg;
An impotent physician
Letting down another patient,
With the excuse of, "Nothing I can do"
To limit my advocacy, my responsibility.
I greet others "like him"
Our systems have forgotten,
Bearing packed plastic bags,
Or pushing shopping carts with all they have.

I wonder what "they" must think
Of this path to paradise we inhabit,
Amidst sparkling new BMWs,
Porsches, and a Lexus;
And rich persons who park cars here,
Sharing the same pavement,
But wearing different vestments
That testament injustice
In a country that permits
A few having much too much,
And too many not near enough,
By any compassionate metric.

There are many such "neighbourhoods"
Where increasingly more
Persons are denied
By the widening divide
Of our unjust class system,
That we idealist students
Were sure we would eradicate,
But is now more firmly fabricated
By the tax breaks we once hated,
And the "trickle-down" philosophies,
Against which we demonstrated
While taking our degrees.

Physicians are prominent pistons
In the contagion of this erosion,
As seen in our frequent
Revolts against taxation.
Yet for those of us whose ethos
Cannot permit that on our watch,
"The system" we fought to fix,
With emphasis on social determinants,
Still exhibits lack of access,
Worsened by the two-tier erosion
Greater profits insist,
But social justice prohibits.

We must assist this new generation
Of socially conscious students
Be the conscience we've forgotten,
And the re-ignition of our once-vision,
To be "a just society,"[294]
First in health and social systems.
We must insist this realization
For the benefit of all Canadians,
So none will be left behind
Beneath a BMW's wheels,
For "None is too many"[295]
For any of us to conceal.

A modified version of the first section of this poem was published as the short story "Homeless Beneath a BMW's Wheels" in the *Canadian Medical Association Journal* in 2020 (Nisker, 2020).

References

Hesse, H. (1968). Beneath the wheel. Farrar, Straus, and Giroux.

Nisker, J. (2020). Homeless beneath a BMW's wheels. *Canadian Medical Association Journal*, 192(28), E815–E816.

---

[294] Trudeau, 1968.

[295] A "senior Canadian official . . . in early 1945, was asked how many Jews would be allowed into Canada after the war. . . . 'None,' he said, 'is too many'" (Troper & Abella, 1982, p. xix).

Troper, H., & Abella, E. (1982). None Is Too Many: Canada and the Jews of Europe, 1933–1948. Lester & Orpen Dennys.

Trudeau, P. (1968, Sept 9). Canada must be a just society. CBC Digital Archives. Video, 2:20. http://www.cbc.ca/player/play/1797431608

Chapter 15

The Injustice of Needing Angelina Jolie

I choose not to keep my story private because there are many women who do not know that they might be living under the shadow of cancer.

—Angelina Jolie, My Medical Choice[296]

The injustice of needing Angelina's story is that Canadian women needed an American to share her story to "know that they might be living under the shadow of cancer,"[297] rather than receiving this caution from Canadian physicians. Angelina's story has prevented many Canadian women from being added to the more than 10,000 Canadian women who died in the decade in which Canada fell behind other "developed" countries in offering BRCA-gene mutation counselling and testing.[298] These Canadian women died because their

---

[296] Jolie, 2013; Angelina was also profiled in *Time Magazine* (Sifferlin, 2014).
[297] Jolie, 2013.
[298] These countries include the Netherlands, the United Kingdom, other European countries, and the United States for many persons, through Medicare, HMOs, and private insurance.

physicians had not been educated to the importance of BRCA gene mutations, and thus lacked the knowledge to offer women at high risk of breast and ovarian cancer the opportunity to choose genetic counselling, testing, MRI surveillance, or prevention strategies.[299] These Canadian women also died because Canadian physicians, as well as Canadian scientists and health-justice advocates,[300] had been impotent in convincing Health Canada, provincial funding agencies, public health agencies, and Canadian health-policy pundits of the importance of publicly funding the genetic counselling, BRCA-gene mutation testing, and increased surveillance required by women at hereditary high risk of early onset breast and ovarian cancer to prevent their deaths.

Angelina Jolie was not aware she carried a BRCA gene mutation until her mother's death prompted her to seek genetic counselling and testing. BRCA gene mutations are autosomal dominant and of high penetrance.[301] "Autosomal dominant" means you only need to inherit one of the two copies of the gene to potentially develop the related condition. High penetrance means if you are one of the 50 percent of your family members who do inherit a copy of the gene, you are very likely to develop the condition. Angelina told the press her story[302] so that women at similar risk of death from breast or ovarian cancer would be aware of their risk, their opportunity for genetic counselling and testing, as well as the opportunity for MRI surveillance that, by detecting cancer at an early stage, improves

---

[299] Nisker, 2013.

[300] I tried to bring awareness of the strong link between BRCA gene mutations and early onset breast cancer to women at risk and the general public through my play *Sarah's Daughters*, which toured across Canada and four other countries; Nisker, 2007, 2012, 2013. See Chapter 10, She Lived with the Knowledge.

[301] Narod, Feunteun, Lynch, Watson, Conway, *et al.*, 1991; Narod, Lynch, Conway, Watson, Feunteun, *et al.*, 1993; Nisker, 2012; See Chapter 2, "You Must Go to Medical School or Hitler Will Have Won"; See Chapter 10, She Lived with the Knowledge.

[302] Jolie, 2013; Sifferlin, 2014.

chances of survival by 70 percent.³⁰³ Angelina's story encouraged women who do test positive for BRCA gene mutations to consider medical or surgical prevention strategies.³⁰⁴ Within two months of Angelina "going public" with her story, there was a tripling of referrals to our Cancer Genetics Clinic;³⁰⁵ many from physicians who had never previously referred. A doubling of referrals was reported in other cancer genetics clinics.³⁰⁶ Without Angelina's story, many of these women would have developed early onset breast or ovarian cancer without appreciating their high risk or their opportunity to mitigate their risk.

Thirty-six years before Angelina told the world her story, Miriam told me her own story.³⁰⁷ Miriam was a 39-year-old woman requesting removal of her ovaries because of her family history of ovarian cancer. My mentor had assured Miriam that ovarian cancer is not hereditary; then asked me to see Miriam and tell him what I thought. Miriam handed me a wrinkled, yellowing sheet of paper with her family tree pencilled on it. Surprisingly, Miriam's family tree indicated her mother, aunt, and a first cousin had died in their forties of ovarian cancer. Miriam's family's ovarian cancer was indeed hereditary. I told Miriam we would remove her ovaries if she was sure she wanted to proceed. As I was leaving her room, a senior medical student was entering to do Miriam's pre-op history and physical. Less than an hour later he ran up to me in a hallway, urging me to come back with him to examine Miriam's breasts. Miriam would not die of ovarian cancer.³⁰⁸

---

³⁰³ Warner, Hill, Causer, Plewes, Jong, et al., 2011.
³⁰⁴ Medical prevention strategies include tamoxifen and other anti-estrogens. However, increased surveillance strategies are more frequently employed, particularly through newer, higher-definition MRI.
³⁰⁵ Joseph, Rab, Panabaker, & Nisker, 2015.
³⁰⁶ Joseph, Rab, Panabaker, & Nisker, 2015; Sifferlin, 2014.
³⁰⁷ See Chapter 6, Miriam.
³⁰⁸ See Chapter 5, Princess Margaret; See Chapter 6, Miriam; See Chapter 10, She Lived with the Knowledge. Miriam's early onset hereditary breast cancer

Twenty-two years before Angelina told the world her story, I was in an airplane, flying to a scientific conference, when a colleague from another university excitedly handed me a folded-open *Lancet*[309] saying, "You've got to read this."[310] The article, written by a Canadian geneticist and epidemiologist, Steven Narod, indicated that the recently cloned BRCA gene mutation[311] was responsible for the majority of deaths from early onset breast cancer.[312] As both physician education and public outreach were urgently required, I began, even on the airplane, to weave creative non-fiction stories of young Canadian women at familial high risk of developing BRCA gene mutation breast cancer into the full-length play *Sarah's Daughters*. The women in *Sarah's Daughters*, and similarly situated women in Canada, had been denied access to genetic counselling and testing because they did not fulfill the strict Canadian qualification of having five family members with breast cancer; five family members many of the women at high risk could not possibly have as their previous generation had perished in the Holocaust.[313]

Sarah's Daughters toured 10 Canadian cities from 2001 to 2003 for public outreach and research purposes.[314] Our analysis of audience members' comments indicated that many Canadian women with high-risk family histories had not heard of BRCA gene

---

death sentence added to my Grandmother's similar death sentence to premonisce my Mother's death sentence.

[309] *The Lancet* is a well-respected British medical journal.

[310] Narod, Feunteun, Lynch, Watson, Conway, *et al.*, 1991.

[311] Narod, Feunteun, Lynch, Watson, Conway, *et al.*, 1991.

[312] See Chapter 10, She Lived with the Knowledge; Narod, Feunteun, Lynch, Watson, Conway, *et al.*, 1991; Narod, Lynch, Conway, Watson, Feunteun, *et al.*, 1993.

[313] See Chapter 2, "You Must Go to Medical School or Hitler Will Have Won"; See Chapter 6, Miriam; See Chapter 10, She Lived with the Knowledge.

[314] The tour was performed with Research Ethics Board approval and was funded by Genome Canada.

mutations,[315] and, perhaps more worrisome, most Canadian physicians had not heard of BRCA gene mutations.[316] *Sarah's Daughters* then toured four other countries for public engagement purposes.

Ten years before Angelina told the world her story, I listened in horror from the back of a social science classroom, as the guest speaker, a well-known Canadian health journalist, confirmed his audience's opinion that the genetics of breast cancer in young women is irrelevant compared with environmental causes.[317] The audience exploded with enthusiastic applause as, in the minds of these feminist scholars and students, genetics was just one more example of medicalization-entrapment.[318]

The following day, I called a close friend, well known internationally for her insightful feminist research and writings. She said it was indeed her understanding that environmental factors are the major causes of breast cancer in young women, and she suggested we together write a book to unpack how social scientists and medical scientists have such differing views regarding the importance of BRCA gene mutations. My friend had several immediate thoughts, including that the Canadian medical profession had a long history of focusing everything on genetics. I agreed with her.[319] Indeed the Canadian medical profession's focus on genetics contributed to Canada's long history of eugenics-based thinking, such as that of Dr. Charles Clarke,[320] as well as eugenics-

---

[315] Nisker, Martin, Bluhm, & Daar, 2006.
[316] Nisker, Martin, Bluhm, & Daar, 2006.
[317] For example, agriculture-chemical spraying and accumulation in the water table.
[318] See Chapter 13, The "Helix of Life" Revisited: DNA in Concrete and Not.
[319] See Chapter 13, The "Helix of Life" Revisited: DNA in Concrete and Not.
[320] See Chapter 13, The "Helix of Life" Revisited: DNA in Concrete and Not; Dr. Clarke was the University of Toronto's first Professor of Psychiatry (1908–1924) and was the first Director of the Toronto Psychiatric Hospital, which, in 1966, was replaced by the Clarke Institute of Psychiatry, named for him (Temerty Faculty of Medicine, 2021). The building is now occupied by

based actions, such as the forced sterilization of "the defective," for which Alberta's physicians were culpable as recently as 1972.[321] The health journalist eventually cut back his catering to believers in the pseudoscience, proclaiming our toxic environment the major cause of early onset breast cancer. However, there was no recognition of the many deaths caused by "BRCA gene mutation-deniers," a charged term I coined to parallel "Holocaust deniers"[322] in recognition of the abhorrent irony that the vast majority of women carrying BRCA gene mutations are of Jewish heritage.

Six years before Angelina told the world her story, I pleaded for a national public education initiative to provide the information required to empower women directly, as relying on their family physicians for this information was not working.[323] Unfortunately, a few years later, it remained clear that Canadian family physicians still lacked the "requisite knowledge" and "lacked time" to discuss genetic testing with their patients,[324] and also "lacked time" to update their patients' family histories.[325] June Carroll, a Canadian family physician, explains these lacks by noting the almost void of cancer-genetics training in medical school and continuing education programs.[326]

Just before Angelina told the world her story, the lack of "requisite" family physicians' knowledge[327] was illustrated in our research with young women at high risk of carrying BRCA gene

---

the Centre for Addiction and Mental Health (CAMH), located just off the University of Toronto's main campus (Dowbiggin, 2005).

[321] Marshall & Robertson, 2006.

[322] Also See Chapter 2, "You Must Go to Medical School or Hitler Will Have Won"; Auschwitz-Birkenau State Museum, 2021.

[323] Nisker, 2007.

[324] Miller, Carroll, Wilson, Bytautas, Allanson, *et al.*, 2010.

[325] Carroll, Rideout, Wilson, Allanson, Blaine, *et al.*, 2009.

[326] Miller, Carroll, Wilson, Bytautas, Allanson, *et al.*, 2010; Carroll, Rideout, Wilson, Allanson, Blaine, *et al.*, 2009.

[327] Miller, Carroll, Wilson, Bytautas, Allanson, *et al.*, 2010; Carroll, Rideout, Wilson, Allanson, Blaine, *et al.*, 2009.

mutations who had not been informed of their high risk until after their breast cancer diagnosis, and subsequent genetics counselling and testing.[328] On being told their BRCA gene mutation test results, the women in our research experienced more intense anger, frustration, grief, and regret than when they had received their breast cancer diagnosis.[329] The women in our research consistently commented that their emotions were triggered by the failure of their physician to make them aware of their risk of carrying a BRCA gene mutation well before they developed breast cancer so they could have considered MRI surveillance or prevention strategies.[330]

---

[328] With research ethics approval, we recruited (through the mailing of letters of information, consent forms, and discussion prompts from their genetics counsellor within their circle of care) women with premenopausal breast cancer who had not been informed about their opportunity to have genetic counselling and testing prior to their mastectomy, and subsequently tested positive for having a BRCA gene mutation. The written comments of these women were transcribed into an electronic format that could be supported by NVivo 9 software for grounded theory analysis. The high response rate indicated how much women wanted to share their stories, despite how difficult it may have been to do so. Despite having a significant family history of breast or ovarian cancer, 50 percent of women had not discussed their family history of breast or ovarian cancer with their family physician prior to their diagnosis, and none of these physicians had discussed genetic counselling and testing. Almost 60 percent of women had thought about developing breast cancer prior to their diagnosis. Over 83 percent of woman wished that they had learned of their risk of developing hereditary cancer from their family physicians. Two-thirds of our research participants also wanted to learn about their risk through public education (Joseph, Rab, Panabaker, & Nisker, 2015; Vanstone, Chow, Lester, Ainsworth, Nisker, *et al.*, 2012).

[329] Joseph, Rab, Panabaker, & Nisker, 2015.

[330] Through their narratives, women volunteering for our research shared coping mechanisms such as networking, being physically active, and staying positive. Women also coped by informing family members about being a BRCA gene mutation carrier and educating them about opportunities for genetic counselling; See Chapter 10, She Lived with the Knowledge; See

In 2021, this injustice continues for the many Canadian women at high risk of early onset breast or ovarian cancer, who remain unaware of their high genetic risk and options to mitigate their risk. Many of these women do not have access to a family physician, the situation for more than 30 percent of persons in Canada.[331] This injustice is furthered when family physicians can be accessed but do not spend adequate time in taking and updating family histories,[332] either because they have too many patients or they have not had sufficient cancer-genetics training.[333]

In 2021, Angelina's American story continues to raise Canadian social-justice questions. Why did Canada need Angelina's story to awake from its delay in preventing deaths from hereditary early onset breast and ovarian cancer, and finally get in step with other "developed" countries[334] in providing education programs to empower women at hereditary genetic risk? Why did Canadian physician organizations need Angelina's story to educate their members? Why did Canadian government funders need Angelina's story to finally institute public funding for BRCA gene mutation testing?

Women at high risk who have not heard Angelina's story, as well as health professionals and students, must hear her story to appreciate the importance of family history in breast and ovarian cancer, and the availability of genetic counselling, genetic testing,

---

Chapter 18, Victor; See Chapter 25, The Arrogance of "But All You Need Is a Good Index Finger"; Joseph, Rab, Panabaker, & Nisker, 2015.

[331] Carroll, Cappelli, Miller, Wilson, Grunfeld, *et al.*, 2008.

[332] Emery, Watson, Rose, & Andermann, 1999; Rich, Burke, Heaton, Haga, Pinsky, *et al.*, 2004; Watson, Shickle, Qureshi, Emery, & Austoker, 1999.

[333] Carroll, Cappelli, Miller, Wilson, Grunfeld, *et al.*, 2008; Carroll, Rideout, Wilson, Allanson, Blaine, *et al.*, 2009; Greendale & Pyeritz, 2001; Ontario Ministry of Health and Long-Term Care, 2000; Watson, Shickle, Qureshi, Emery, & Austoker, 1999.

[334] For example, educational programs are available in the Netherlands, the United Kingdom, other European countries, and even in the United States through Medicare and HMOs.

MRI surveillance,[335] and prevention strategies. Angelina Jolie concluded her story with, "Life comes with many challenges. The ones that should not scare us are the ones we can take on and take control of."[336] We still need Angelina's story in Canada so that women can "take control of" their genetic risk of breast and ovarian cancer.

I want to encourage every woman, especially if you have a family history of breast or ovarian cancer, to seek out the information and medical experts who can help you through this aspect of your life, and to make your own informed choices.

—Angelina Jolie, My Medical Choice[337]

An earlier version of "The Injustice of Needing Angelina Jolie" was published in the Journal of Obstetrics and Gynaecology Canada.[338]

References

Auschwitz-Birkenau State Museum. (2021). *Deniers in different countries.* http://auschwitz.org/en/history/holocaust-denial/deniers-in-different-countries

Carroll, J. C., Cappelli, M., Miller, F., Wilson, B. J., Grunfeld, E., Peeters, C., Hunter, A. G., Gilpin, C., & Prakash, P. (2008). Genetic services for hereditary breast/ovarian and colorectal

---

[335] Being identified as a BRCA gene mutation carrier qualifies women for breast MRI surveillance, starting as young as age 29 (Warner, Hill, Causer, Plewes, Jong, et al., 2011); Jolie, 2013.
[336] Jolie, 2013.
[337] Jolie, 2013.
[338] Nisker, 2013.

cancers—Physicians' awareness, use and satisfaction. *Community Genetics, 11*(1), 43–51.

Carroll, J. C., Rideout, A. L., Wilson, B. J., Allanson, J. M., Blaine, S. M., Esplen, M. J., Farrell, S. A., Graham, G. E., MacKenzie, J., Meschino, W., Miller, F., Prakash, P., Shuman, C., Summers, A., & Taylor, S. (2009). Genetic education for primary care providers: Improving attitudes, knowledge, and confidence. *Canadian Family Physician 55*(12), e92–e99. https://www.ncbi.nlm.nih.gov/pubmed/20008584

Dowbiggin, I. (2005). Clarke, Charles Kirk. In *Dictionary of Canadian biography*, vol. 15. University of Toronto/Université Laval. http://www.biographi.ca/en/bio/clarke_charles_kirk_15E.html

Emery, J., Watson, E., Rose, P., & Andermann, A. (1999). A systematic review of the literature exploring the role of primary care in genetic services. *Family Practice, 16*(4), 426–445.

Greendale, K., & Pyeritz, R. E. (2001). Empowering primary care health professionals in medical genetics: How soon? How fast? How far? *American Journal of Medical Genetics, 106*(3), 223–232.

Jolie, A. (2013, May 14). *My medical choice.* New York Times. https://www.nytimes.com/2013/05/14/opinion/my-medical-choice.html

Joseph, M., Rab, F., Panabaker, K., & Nisker, J. (2015). Feelings of women with strong family histories who subsequent to their breast cancer diagnosis tested BRCA positive. *International Journal of Gynecological Cancer 25*(4), 584–592.

Marshall, T., & Robertson, G. (2006). Eugenics in Canada. In *The Canadian encyclopedia*. Historica Canada. https://www.thecanadianencyclopedia.ca/en/article/eugenics

Miller, F. A., Carroll, J. C., Wilson, B. J., Bytautas, J. P., Allanson, J., Cappelli, M., de Laat, S., & Saibil, F. (2010). The primary care physician role in cancer genetics: A qualitative study of patient experience. *Family Practice, 27*(5), 563–569.

Narod, S., Lynch, H., Conway, T., Watson, P., Feunteun, J., & Lenoir, G. (1993). Increasing incidence of breast cancer in family with BRCA1 mutation. *Lancet, 341*(8852), 1101–1102.

Narod, S. A., Feunteun, J., Lynch, H. T., Watson, P., Conway, T., Lynch, J., & Lenoir, G. M. (1991). Familial breast-ovarian cancer locus on chromosome 17q12-q23. *Lancet, 338*(8759), 82–83.

Nisker, J. (2012). *From Calcedonies to Orchids: Plays promoting humanity in health policy.* Iguana Books.

Nisker, J. (2013). A public health education initiative for women with a family history of breast/ovarian cancer: Why did it take Angelina Jolie? *Journal of Obstetrics and Gynaecology Canada, 35*(8), 689–691.

Nisker, J., Martin, D. K., Bluhm, R., & Daar, A. S. (2006). Theatre as a public engagement tool for health-policy development. *Health Policy, 78*(2–3), 258–271.

Nisker, J. A. (2007). The need for public education: "Surveillance and risk reduction strategies" for women at risk for carrying BRCA gene mutations. *Journal of Obstetrics and Gynaecology Canada, 29*(6), 510–511.

Ontario Ministry of Health and Long-Term Care. (2000). Provincial predictive genetic testing service for hereditary breast, ovarian and colon cancers (Bulletin 4352). http://www.health.gov.on.ca/en/pro/programs/ohip/bulletins/4352/bul4352.aspx

Rich, E. C., Burke, W., Heaton, C. J., Haga, S., Pinsky, L., Short, M. P., & Acheson, L. (2004). Reconsidering the family history in primary care. *Journal of General Internal Medicine, 19*(3), 273–280.

Sifferlin, A. (2014, September 2). Angelina Jolie's surgery may have doubled genetic testing rates at one clinic. *Time Magazine.* https://time.com/3256718/angelina-jolie-genetic-testing/

Temerty Faculty of Medicine. (2021). Our history. Department of Psychiatry, University of Toronto, https://www.psychiatry.utoronto.ca/our-history

Vanstone, M., Chow, W., Lester, L., Ainsworth, P., Nisker, J., & Brackstone, M. (2012). Recognizing BRCA gene mutation risk subsequent to breast cancer diagnosis in southwestern Ontario. *Canadian Family Physician*, *58*(5), e258–e266. https://www.ncbi.nlm.nih.gov/pubmed/22734169

Warner, E., Hill, K., Causer, P., Plewes, D., Jong, R., Yaffe, M., Foulkes, W. D., Ghadirian, P., Lynch, H., Couch, F., Wong, J., Wright, F., Sun, P., & Narod, S. A. (2011). Prospective study of breast cancer incidence in women with a BRCA1 or BRCA2 mutation under surveillance with and without magnetic resonance imaging. *Journal of Clinical Oncology*, *29*(13), 1664–1669.

Watson, E. K., Shickle, D., Qureshi, N., Emery, J., & Austoker, J. (1999). The "new genetics" and primary care: GPs' views on their role and their educational needs. *Family Practice*, *16*(4), 420–425.

Chapter 16

A Brief and Personal History[339] of "What's in a Name"

What's in a name? That which we call a rose by any other name would smell as sweet.

—Juliet Capulet in William Shakespeare's *Romeo and Juliet*[340]

Although Juliet Capulet's claim of "What's in a name?" may be correct when applied to her family name and that of her suitor, Romeo Montague,[341] "that which we call" embryos and procedures in reproductive genetics often smell sweet because the names were created to perfume not-so-sweet-smelling practices. Over the past 30 years, reproductive-genetics scientists and clinicians, myself included, have used perfumed names, knowingly and unknowingly, to make our research smell sweet for research ethics boards (REBs), research-funding agency reviewers, government regulators, hospital

---

[339] I borrow "A Brief History" from Stephen Hawking's groundbreaking book, *A Brief History of Time: From the Big Bang to Black Holes* (Hawking, 1988).
[340] Shakespeare, 1734.
[341] Shakespeare, 1734.

administrators, and the general public. Scientists' and clinicians' use of sweet-smelling names promotes research innovation, clinical acceptance, and public reassurance, not to mention academic acclaim and financial gain. However, sweet-smelling names also promote deception, not only of REBs, research funders, policy makers, and the public, but of the women supposedly participating in informed-choice processes impacting their health. Sweet-smelling names also have social impacts, including on our perceptions of persons with disabilities.[342]

The sweet-smelling names in reproductive genetics explored here include "pre-embryo," pre-implantation genetic "diagnosis," "normal" embryo, "unsuitable" embryo, "healthy" embryo, pre-implantation genetic "testing," non-invasive prenatal "testing," "donation," and most recently "mitochondrial replacement therapy," a sweet-smelling name for germ-line nuclear transfer prohibited in most countries in their anti–reproductive cloning legislation. As Co-Chair of Health Canada's Advisory Committee on Reproductive and Genetic Technology for our *Assisted Human Reproduction Act* (2004), I observed a clear consensus of the panel of scientists, clinicians, ethicists, legal scholars, and members of the general public on the prohibition of reproductive cloning.

The sweet-smelling name "pre-embryo" was created in the 1980s to describe human embryos fewer than 14 days post fertilization.[343] The name "pre-embryo" had not previously been used in animal or human embryology or in cell biology,[344] but was thought to be sweeter smelling than the name "embryo" for the noses of REB members and national policy developers. I was guilty of using the perfumed name "pre-embryo," knowingly or unknowingly, in my REB submissions for the mouse and human embryo research[345] that we hoped would

---

[342] Joseph, Saravanabavan, & Nisker, 2018; Mykitiuk & Nisker, 2010; Nisker, 2001, 2012.
[343] Colomer & Pastor, 2012.
[344] Colomer & Pastor, 2012.
[345] Gore-Langton, Nisker, & Natale, 1993; Nisker, 1992; Nisker & Gore-Langton, 1995.

contribute to the development of pre-implantation genetic diagnosis (PGD).[346] Originally for PGD, cells of eight-cell "pre-embryos" were removed through a glass pipette,[347] their DNA multiplied millions of fold through the polymerase chain reaction (PCR),[348] then their genetics screened (rather than tested, as will be discussed in the next section). Only "pre-embryos" not having the concerning genetic structure that promoted their biopsy were implanted in the woman's uterus or cryopreserved for later implantation. PGD was designed as a new technology we hoped would obviate the risks of amniocentesis in women undergoing IVF.[349] The use of the name "pre-embryo" ceased when PGD was fully accepted by regulators, and somewhat accepted by the general public.[350]

I had also sweetened the smell of our research by emphasizing that we were experimenting on the "multipronucleate" "pre-embryos" that occur in 15 percent of human fertilizations.[351] We thought "multipronucleate" "pre-embryos," having 23 extra chromosomes and thus not a human karyotype,[352] would not be considered human embryos. However, by 1993, I was concerned with the concept of chromosome number determining our humanness and the danger this concept poses to persons whose chromosome number varies from 46. These include persons with one extra chromosome, such as persons with Down syndrome or Klinefelter syndrome, as well as persons with one fewer chromosome, such as persons with Turner syndrome.

Pre-implantation genetic "diagnosis" is itself a perfumed name, as PGD is generally used for embryo *screening* not "diagnosis."

---

[346] Handyside, Kontogianni, Hardy, & Winston, 1990; Handyside, Pattinson, Penketh, Delhanty, Winston, *et al.*, 1989; Nisker, 1992; Nisker & Gore-Langton, 1995.

[347] Handyside, Pattinson, Penketh, Delhanty, Winston, *et al.*, 1989; Nisker & Gore-Langton, 1995.

[348] Mullis, Faloona, Scharf, Saiki, Horn, *et al.*, 1986.

[349] Nisker, 2001; Nisker & Gore-Langton, 1995.

[350] Colomer & Pastor, 2012.

[351] Gore-Langton, Nisker, & Natale, 1993; Nisker & Gore-Langton, 1995.

[352] *Karyotype* comes from the Greek word for "kernel, seed, or nucleus."

However, "diagnosis" sprays a socially acceptable scent on embryo *screening*, which in essence is *screening out* and possesses a eugenic odour. The perfumed name "diagnosis" also suggests a potential therapeutic purpose, which is not possible for the embryo "diagnosed," although emerging gene-editing technologies are on the horizon, albeit with ethics, legal, social, and policy concerns.

I did use *screening* in the name for our PGD laboratory, the "Early Pre-Implantation Cell Screening (EPiCS) Laboratory," but added "Early" and "Cell" to sweeten the smell of *screenings* for our University's REB, research funders, hospital administrators, and the general public. The perfumed-name "early" suggested the eight-"cell" human "pre-embryos" we were using in our research had not yet developed into true human embryos. The perfumed-name "cell" camouflaged the reality that each of the two blastomeres[353] we were removing from eight-"cell" "pre-embryos" was totipotential and thus could develop into a child. I even put up EPiCS posters along the hospital corridor walls outside our laboratory to illustrate that we were simply removing two "cells" from eight-"cell" "pre-embryos" to reassure passersby that we were not evil scientists, manipulating human embryos, like in Ira Levin's *The Boys from Brazil*[354] or the newspaper images of reproductive-cloning, evil-looking scientists of that time that caused the legislations prohibiting reproductive cloning to be enacted.

In 1993, after a leak to the press and subsequent requests to come to our Unit for PGD to select XY embryos for "family completion" (55 of 74 phone calls), I closed the EPiCs Laboratory, returned the research funding, and began exploring the ethics, legal, social, and policy issues emanating from PGD research and its potential clinical practice. This decision was partly a *mea culpa* to Professor Abby Lippman (McGill University), Professor Susan Sherwin (Dalhousie

---

[353] Handyside, Kontogianni, Hardy, & Winston, 1990; Handyside, Pattinson, Penketh, Delhanty, Winston, *et al.*, 1989.
[354] Levin, 1976.

University), and other feminist friends who had warned me where my research would lead. My *mea culpa* was not completed until 1995, when I completed my full-length play *Orchids*.[355]

Subsequent to our basic science research on PGD, I became concerned with "what's in a name" like "normal" (and thus "abnormal") when referring to IVF embryos that had undergone PGD, and the impact these names would have on our perceptions of children and adults with genetic conditions. In 1995, to share my concerns with policy makers, scientists, clinicians, and the general public, I wrote the two-hour play *Orchids*. *Orchids* explores the concept of "normal" through the interwoven stories of two women who become friends in a series of IVF Unit waiting-room waits; one woman undergoing IVF for the purpose of bypassing her blocked fallopian tubes, the other for the purpose of PGD to ensure her child will not have the same "horrible" genetic condition as her brother.

As *Orchids* is performed, the audience learns that the woman with blocked fallopian tubes has the same genetic condition as her waiting-room friend's brother but does not want PGD, because she always thought that she was "normal." The tensions that develop between the central characters bring the audience to contemplation of what's in a name like "normal." The cast also includes two physicians, one enamoured with the new capacities of PGD and one worried where PGD may lead, as well as a chorus of IVF lab technicians proud to be "embryo engineers."[356]

The second cross-Canada tour of *Orchids* in 2008, funded as both a public engagement and research project by Canadian Institutes of Health Research and Health Canada, was viewed by 741 audience members.[357] In our empirical analysis of the comments of audience members *qua* research participants following each performance (in large groups, small focus groups, and written surveys), we found that audience members expressed hesitancy to

---

[355] Nisker, 2001, 2012.
[356] Nisker, 2001, 2012.
[357] Cox, Kazubowski-Houston, & Nisker, 2009.

"draw a line between acceptable and unacceptable indications for PGD" because of their concerns regarding the implications of such a line. In addition, audience members were concerned that neither the choices of the medical establishment nor individual women undergoing IVF should determine the acceptable indications for PGD. The audience members indeed brought attention to how collective choices have individual effects and individual choices have collective effects.

In subsequent years, audience members have expressed similar concerns following other productions of *Orchids* and in discussions following my public engagements on the implications of "normal" and "abnormal" when referring to the genetics of embryos, children, and adults. In 2012, in a week-long run of *Orchids* produced in Norfolk, Virginia, the director chose to employ an all-Black, all-women cast to challenge audiences to go deeper in considering the implications of being different from what is generally expected.

The excitement over the potentials of stem cell research resulted in the sweet-smelling name "unsuitable" for embryos that morphologically might be considered less "suitable" to achieve pregnancy than other embryos in the Petri dish, and thus could be "donated" "fresh" to stem cell research. The name "unsuitable" embryo is a sweet-smelling name because it allows the embryo it describes to be discarded from its cohort, rather than being cryopreserved for the woman's later reproductive use to obviate additional IVF cycles. The reason for the creation of the sweet-smelling name "unsuitable" was the eagerness of stem cell scientists to access "fresh" human embryos, as they believed "fresh" embryos were more "suitable" for their research than the many cryopreserved embryos that had been altruistically donated and stored for research purposes. In our Unit, over 500 IVF embryos had been cryopreserved with the designation of "donation to research" when no longer required for reproductive purposes.[358]

---

[358] Newton, McDermid, Tekpetey, & Tummon, 2003.

The name "unsuitable" embryo, for embryos considered to be less "suitable" to result in a pregnancy, was based on the primitive microscope grading scales of that time. Prior to stem cell research, cryopreservation of all embryos not transferred "fresh" was the standard of care except for women who could not afford the cryopreservation fee or had a moral problem with cryopreservation. In the era of stem cell research, embryos labelled "unsuitable" were encouraged to be "donated" "fresh" to stem cell research rather than being transferred "fresh" to the uterus of the woman who created them or cryopreserved for her later reproductive use.

As stem cell research took off, the microscopic designation of "unsuitable" became more frequently applied to IVF embryos, even though an embryo's potential to develop into a child could not be ruled out at that time prior to the cessation of embryo cell division or embryo degeneration. In Australia, where prior to stem cell research, no IVF embryo was considered "unsuitable," one Australian IVF clinician commented to me that "Even the ugliest embryos can become children," and another said, "Why not give them all a chance?" However, the preference of Australian stem cell researchers to use "fresh" embryos promoted a National Health and Medical Research Council Guideline containing microscopic criteria under which embryos could be considered "unsuitable" and could be donated as "fresh" embryos to stem cell research.[359]

Another deeply problematic sweet-smelling name for IVF embryos is "healthy," which implies that except for particularly favourable characteristics, determined by genetic or non-genetic means, there also exist "unhealthy" embryos. I was a Co-Principal Investigator in a large multidisciplinary, multicentre Canadian Institutes of Health Research (CIHR) team grant with the sweet-smelling working title "The Healthy Embryo." I should have noticed the inappropriateness of the name "healthy" when preceding "embryo," as I was leader of the Ethics, Law, and Society research

---

[359] Mykitiuk & Nisker, 2010.

team, and, as noted above, had previously researched the inappropriateness of the names "normal" and "unsuitable" when applied to embryos as noted above. The inappropriateness of the name "healthy embryo" was brought to my attention by the legal scholar on our team, Professor Roxanne Mykitiuk, whose research focuses on discrimination against persons with disabilities.

Professor Mykitiuk and I then began researching the ethics, legal, and social implications of the concept of a "healthy embryo" and found that "both social and biomedical determinants of embryo health exist within social contexts and have normative and clinical implications."[360] We also found that, similar to my previous research on "normal" embryos, our perceptions of "healthy" children influence our perceptions of "healthy" embryos. We concluded that future determinations regarding "healthy" embryos would promote problematic new characterizations of the health of children and adults.[361]

To further explore the impact of the names "healthy" and "unhealthy" embryos on children and adults with disabilities, genetic- or non–genetic-related, I used some of the CIHR funding to host an international conference on the concept of a "healthy embryo," inviting researchers, clinicians, and scholars from several countries. The conference resulted in the book *The "Healthy" Embryo: Social, Biomedical, Legal and Philosophical Perspectives*, published by Cambridge University Press.[362]

The burgeoning of new reproductive-genetics capacities for the designation of "healthy embryos" will in the near future include modification of the genetics of embryos deemed "unhealthy" through new gene-editing capacities. These new reproductive-genetics capacities will require extensive ethics, legal, and social research, as well as national and international regulations.

Using the sweet smell of "testing" in pre-implantation genetic "testing" (PGT) again perfumes the name "screening," and thus the

---

[360] Mykitiuk & Nisker, 2010, p. 128.
[361] Mykitiuk & Nisker, 2010, p. 128.
[362] Nisker, Baylis, Karpin, McLeod, & Mykitiuk, 2010.

eugenic smell of "screening out." Pre-implantation genetic "screening" (PGS) is indeed the appropriate name except when "testing" for specific hereditary conditions, such as monogenic conditions (PGT-M); or when "testing" for specific "aneuploidies" (PGT-A) in the embryos of women of advanced reproductive age (such an extra chromosome 21 or 18) or when testing for any aneuploidy in the embryos of women with a history of recurrent pregnancy loss.

Although some studies indicate higher singleton birth rates after PGS is used to screen embryos for single embryo transfer (SET) to prevent multiple pregnancies, debate continues as to whether PGS (including PGT-A) actually does increase the pregnancy rate. Similarly, although PGS was thought to be useful in preventing recurrent pregnancy loss, studies comparing pregnancy rates of IVF cycles using PGT-A to IVF cycles without PGT-A have not confirmed this to be the case. PGS, if proven to promote a higher pregnancy rate with SET or a decreased pregnancy-loss rate, would be important in decreasing the physical and psychological harms of IVF and recurrent pregnancy loss.

Again, "what's in a name," but it seems to me that "screening," as in PGS, is the appropriate name in most circumstances of assessing embryos, and indeed this name is increasingly being used. However, in countries where IVF is not publicly funded, PGS may add an extra cost to the woman (or couple), and I contend that it is inappropriate that women should be charged extra to prevent miscarriages if indeed PGS can assist with miscarriage prevention, but particularly when the scientific evidence is not yet convincing. The Practice Committee of the American Society for Reproductive Medicine cautions PGS should not be routinely employed in all IVF cycles, as is becoming routine.[363]

Another sweet-smelling name in reproductive genetics is "non-invasive prenatal testing" (NIPT), which is neither "non-invasive"

---

[363] Practice Committee of Society for Assisted Reproductive Technology; Practice Committee of American Society for Reproductive Medicine, 2008.

nor, in the vast majority of circumstances, "testing." NIPT was developed as a technology to test the DNA in fetal cells sequestered from a pregnant woman's blood.[364] NIPT was quickly commercialized by four biotech companies to be an over-the-counter "testing" kit[365] that a pregnant woman, as early as at eight weeks' gestation,[366] can purchase a kit, then provide a blood sample (usually at a private laboratory) and send it with a hundred dollars to a biotech company chosen from online or pamphlet advertisements.[367] At the biotech company, the embryo cells are "screened" for genetic "abnormalities"[368] through the use of new microarray[369] and exome sequencing[370] technologies. The results of the "screen" are then mailed directly to the woman, whose physician or midwife likely has not yet been involved; nor has any genetic counselling occurred.[371]

The "N" and "I" in NIPT emphasize the sweet smell of "non-invasive," spraying a scent of safety on a technology that can be quite invasive, particularly psychologically, for the many women who receive a confusing and anxiety-provoking genetic print-out directly from the biotech company, without having previous genetic counselling or the opportunity to have rapid access to genetic counselling.[372] The "T" in NIPT represents the sweeter-smelling name "testing" to again perfume the less sweet-smelling name "screening," which, as discussed above, in essence means "screening out" and has a eugenic scent. However, the NIPT "kits"

---

[364] Farra, Choucair, & Awwad, 2018; Lo, Corbetta, Chamberlain, Rai, Sargent, *et al.*, 1997; Renga, 2018.
[365] Vanstone, King, deVrijer, & Nisker, 2014.
[366] Farra, Choucair, & Awwad, 2018; Lo, Corbetta, Chamberlain, Rai, Sargent, *et al.*, 1997; Renga, 2018.
[367] Vanstone, King, deVrijer, & Nisker, 2014.
[368] Wright & Burton, 2009; Wright, Wei, Higgins, & Sagoo, 2012.
[369] Van den Veyver, Patel, Shaw, Pursley, Kang, *et al.*, 2009.
[370] Drury, Williams, Trump, Boustred, Gosgene, *et al.*, 2015.
[371] Vanstone, King, deVrijer, & Nisker, 2014; Vanstone, Yacoub, Winsor, Giacomini, & Nisker, 2015.
[372] Vanstone, King, deVrijer, & Nisker, 2014.

are in fact designed for "screening," as the vast majority of pregnant women purchasing these "kits" have no specific genetic risks for which to "test," unless the genetic risk is related to their age or to having a fetus of the unwanted sex[373] or, more recently, Rh status.[374] "Testing" is the only appropriate name when NIPT is used as a step before amniocentesis and after genetic counselling regarding a particular genetic condition or a "non-reassuring" integrated prenatal "screen" (IPS).[375] The biotech companies that promote the sweet smell of NIPT of course envisage a much larger market for their "kits" than only the women for whom professional bodies recommend their use.[376]

Although I was following the basic science research studying DNA in fetal cells sequestered from a pregnant woman's blood, I first learned that NIPT was in clinical use from an obstetrics resident's grand rounds presentation on NIPT. She described in her talk that NIPT "kits" were available in our prenatal clinic. Subsequent to my disbelief and "please show me" after her presentation, she led me upstairs to our prenatal clinic and pointed to a pyramid pile of NIPT "kits" that a drug rep had stacked on a table in the waiting room. Our investigation determined these "kits" were stacked without knowledge of our Department's faculty or staff, or of the hospital's administration.

The sweet-smelling name "non-invasive" conceals the numerous and not-so-sweet-smelling ethical issues that arise when marketing supersedes genetic counselling, including lack of informed choice and the receiving of personal "health" information from a biotech company rather than from a health professional, particularly without rapid access to genetic counselling to interpret the complex information received; regarding not only the genetics of the embryo

---

[373] Vanstone, King, deVrijer, & Nisker, 2014.
[374] Fung & Eason, 2018.
[375] Vanstone, King, deVrijer, & Nisker, 2014.
[376] Langlois, Brock, Wilson, Audibert, Brock, *et al.*, 2013; Vanstone, King, deVrijer, & Nisker, 2014.

but the genetics of the pregnant woman and predictions about her future "health." The information received from the biotech company is not only complex but frequently misleading.[377]

There are also problematic ethics, legal, social, and policy issues regarding biotech companies promoting their products in publicly funded institutions such as prenatal clinics in hospitals, but also in other well-respected settings such as physicians' offices and pharmacies. The existence of these displays promoting NIPT "kits" in respected locations promotes the perception that the NIPT "kits" are approved by professional organizations, which again is not the case except in particular situations as indicated by national professional organization guidelines and policy statements.

"Donation" is another sweet-smelling name in reproductive genetics, whether the "donation" is of oocytes, sperm, or embryos. In reproductive genetics, sperm or oocyte "donation" is generally chosen to avoid a prospective child inheriting the genetic condition carried by either one prospective parent (autosomal dominant or X-linked recessive) or both (autosomal recessive), while still allowing a genetic relationship to one of the parents. The sweet smell of "donation" covers the reality that the sperm and oocytes are generally *purchased* in most countries, though the altruistic donation of sperm and oocytes does occur. As the production and sale of sperm for "donation" has much less risk of physical harm than oocyte donation, its financial compensation is much less.

For the process of oocyte "donation," a usually socio-economically disadvantaged woman is prescribed a sequence of drugs to stimulate her ovaries to produce multiple oocytes that a physician will surgically remove to be sold by an "oocyte broker" to a purchasing woman (or couple). As the selling of oocytes is illegal in Canada under the commodification section of the *Assisted Human Reproduction Act*[378] and also illegal in other countries with

---

[377] Vanstone, King, deVrijer, & Nisker, 2014.
[378] Government of Canada, 2004.

similar legislation, the sweet-smelling name "donation" sprays the scent of altruism analogous to that of blood donation, kidney donation, other organ or tissue donation, or even anonymous financial donation.

The perfume of "donation" is also essential to cover the untoward scent of physicians prescribing drugs and performing surgery with no health benefit to their oocyte "donor" patient, rather the potential for serious physical harms, such as ovarian hyperstimulation syndrome, and also psychological harms. Physicians who participate in this practice, I contend, break their Hippocratic Oath, as rather than having the best interest of their oocyte "donor" patients (again, usually a socio-economically disadvantaged woman), they have the best interest of the purchasing woman (or couple), who pays the physicians a large sum of money under the sweet-smelling name "donation."

The sweet smell of oocyte "donation" has been working so well in Canada that, with the exception of that of one "oocyte broker,"[379] no criminal sanction has been enforced under our *Assisted Human Reproduction Act*.[380]

"Mitochondrial replacement therapy" is not only the newest sweet-smelling name in reproductive genetics, it is the most deceiving, and I contend the most problematic. For this supposed "mitochondrial replacement," the nucleus of an IVF oocyte from a woman who wants to be a genetically related parent is inserted into the enucleated oocyte of a "donor," a woman, hired to undergo the risks of IVF to provide oocytes containing cytoplasm with supposedly "healthy" mitochondria.[381] "Mitochondrial replacement" thus is not replacement of the mitochondria at all; rather, it is a sweet-smelling name for germ-line nuclear transfer and thus reproductive cloning.[382] Further, the sweet-smelling name "therapy" soothes people into

---

[379] Baylis, 2012; Blackwell, 2012.
[380] Government of Canada, 2004.
[381] Baylis, 2013; Lane & Nisker, 2016; Nisker, 2015.
[382] Baylis, 2013; Lane & Nisker, 2016; Nisker, 2015.

thinking that the procedure must be good because it is therapeutic, while the foul-smelling name "reproductive cloning" again surfaces images of evil scientists[383] that caused the legislation prohibiting reproductive cloning to be enacted.

It was necessary to invent the perfumed name "mitochondrial replacement therapy" to circumvent the prohibition against germ-line nuclear transfer in the reproductive cloning sections of the legislation of all countries where laws governing assisted reproductive and human embryo research exist,[384] and even in the standalone *Human Cloning Prohibition Act*[385] in the United States. Surprisingly, the United Kingdom accepted the sweet smell of "mitochondrial replacement therapy" in 2015, although legislation against the fouler-smelling "reproductive cloning" continues to exist there.[386] Part of my surprise stems from the United Kingdom's history of leading the way in the regulations restraining reproductive-genetics technology. It is likely that the UK's Scientific and Clinical Advances Advisory Committee (SCAAC) of the Human Fertilisation and Embryology Authority (HFEA) felt by approving "mitochondrial replacement therapy" with tight restrictions and emphasizing that "mitochondrial replacement therapy" would be used only for a few citizens a year, the United Kingdom would continue to lead the way in regulating reproductive genetics. However, by accepting the sweet-smelling name "mitochondrial replacement therapy," the HFEA diminished the many ethical and social concerns of reproductive cloning that

---

[383] Charismatic French chemist cuts chic figure as she promotes human cloning for Raelian sect, 2003; Cohen, 1998; Gibbs, 2001; Higgins, 2001; Levin, 1976.

[384] *Act Respecting Research on in Vitro Embryos*, 2003; Government of Canada, 2004; *Human Fertilisation and Embryology Act 1990*, 1990; *Law No. 94-654 of July 29, 1994 Relating to the Donation and Use of Elements and Products of the Human Body, Medically Assisted Procreation and Prenatal Diagnosis*, 1994; *Prohibition of Human Cloning for Reproduction Act 2002*, 2002.

[385] *Human Cloning Prohibition Act of 2003*, 2003.

[386] *Human Fertilisation and Embryology Act 1990*, 1990.

caused anti-cloning legislation to be enacted in the United Kingdom and many other countries.[387]

The HFEA's decision to permit the sweet-smelling "mitochondrial replacement therapy" did not seem to worry about the not-so-sweet-smelling potential physical[388] and psychological harms[389] to the woman hired to be the oocyte "donor" and undergo the IVF drugs and surgery to provide oocytes for the woman (or couple) choosing "mitochondrial replacement therapy" (as described above). The HFEA SCAAC was also not concerned with the sweet-smelling name "therapy" as used in "mitochondrial replacement therapy," even though no actual "therapy" is taking place.[390]

The main purpose of the promoters of "mitochondrial replacement therapy" likely goes beyond the HFEA SCAAC's facilitation of a few mother–child genetic relationships a year; rather, it extends to the hope that after legislative approval of this enterprise it will be possible to pursue a wide variety of research endeavours in which germ-line nuclear transfer in human embryos can be used. Again, as I have suggested throughout this narrative exploration, the promoters of this new reproductive-genetics technology, similar to the promoters of previous reproductive-genetics technologies, cannot vacate their interest, either in academic acclaim or in reproductive-genetics technologies that are employed more frequently and are more profitable.

---

[387] *Act Respecting Research on in Vitro Embryos*, 2003; Government of Canada, 2004; *Human Cloning Prohibition Act of 2003*, 2003; *Human Fertilisation and Embryology Act 1990*, 1990; *Law No. 94-654 of July 29, 1994 Relating to the Donation and Use of Elements and Products of the Human Body, Medically Assisted Procreation and Prenatal Diagnosis*, 1994; *Prohibition of Human Cloning for Reproduction Act 2002*, 2002.

[388] Abramov, Elchalal, & Schenker, 1999; Asch, Li, Balmaceda, Weckstein, & Stone, 1991; Balen, 2005; Klemetti, Sevon, Gissler, & Hemminki, 2005; Practice Committee of the American Society for Reproductive Medicine, 2003; Schenker & Ezra, 1994; Swanton, Storey, McVeigh, & Child, 2010.

[389] Randal, 2004.

[390] Nisker, 2015.

In conclusion, to respond to Juliet Capulet's claim, "What's in a name? That which we call a rose by any other name would smell as sweet,"[391] I contend there is much in names in reproductive genetics, and much in the motivations of those who create the sweet-smelling names. Sweet-smelling names often camouflage not-so-sweet-smelling practices from the noses of research ethics board members, research-funding agency members, government regulators, hospital administrators, and the general public. Sweet-smelling names are used to promote research innovation, clinical acceptance, and public reassurance, but sweet-smelling names also promote deception, not only of the above persons but, more important, for the women undergoing informed-choice processes as they contemplate reproductive-genetics procedures that could have major impacts on their health. This deception in reproductive genetics also has social impacts, including our perceptions of persons with disabilities.

Although the perfumed name "pre-embryo" is no longer in use, and the concepts of an "unsuitable" embryo, a "normal" embryo, and a "healthy" embryo have been challenged, non-invasive prenatal "testing" and "mitochondrial replacement therapy" have begun to infiltrate medical parlance. In order for informed choices to occur for women, for research ethics boards and for other regulators, it is essential that we researchers and clinicians, as well as the general public and our professional organizations, be engaged in considering the real meaning of sweet-smelling names in reproductive genetics.

A version of "A Brief and Personal History of 'What's in a Name'" was published in *Medical Humanities* in June 2021.[392]

---

[391] Shakespeare, 1734.
[392] Nisker, 2021.

References

Abramov, Y., Elchalal, U., & Schenker, J. G. (1999). Severe OHSS: An "epidemic" of severe OHSS: A price we have to pay? *Human Reproduction, 14*(9), 2181–2183.

Act Respecting Research on in Vitro Embryos. (2003). The Federal Government, Belgium. https://studylibfr.com/doc/584210/11-mai-2003.---loi-relative-%C3%A0-la-recherche-sur-les-embryo...

Asch, R. H., Li, H. P., Balmaceda, J. P., Weckstein, L. N., & Stone, S. C. (1991). Severe ovarian hyperstimulation syndrome in assisted reproductive technology: Definition of high risk groups. *Human Reproduction, 6*(10), 1395–1399.

Balen, A. (2005). *Ovarian hyperstimulation syndrome—A short report for the HFEA.* http://data.parliament.uk/DepositedPapers/Files/DEP2008-2052/DEP2008-2052.pdf

Baylis, F. (2012). The demise of Assisted Human Reproduction Canada. *Journal of Obstetrics and Gynaecology Canada, 34*(6), 511–513. https://www.ncbi.nlm.nih.gov/pubmed/22673165

Baylis, F. (2013). The ethics of creating children with three genetic parents. *Reproductive Biomedicine Online, 26*(6), 531–534.

Blackwell, T. (2012, March 1). Fertility consultant at centre of RCMP raid in the dark about reason for investigation: Lawyer. *National Post.* https://nationalpost.com/news/canada/fertility-consultant-at-centre-of-rcmp-raid-in-the-dark-about-reason-for-investigation-lawyer

Charismatic French chemist cuts chic figure as she promotes human cloning for Raelian sect. (2003, January 4). *Globe and Mail.* https://www.theglobeandmail.com/news/world/charismatic-french-chemist-cuts-chic-figure-as-she-promotes-human-cloning-for-raelian-sect/article22392940/

Cohen, P. (1998, January 17). Crossing the line—Richard Seed may not win a place in history for cloning humans, but someone probably will. *New Scientist.* https://www.newscientist.com/article/mg15721170-200-crossing-the-line-richard-seed-may-not-win-a-place-in-history-for-cloning-humans-but-someone-probably-will/

Colomer, M. F., & Pastor, L. M. (2012). The preembryo's short lifetime. The history of a word. *Cuadernos de Bioética*, *23*(79), 677–694.

Cox, S. M., Kazubowski-Houston, M., & Nisker, J. (2009). Genetics on stage: Public engagement in health policy development on preimplantation genetic diagnosis. *Social Science & Medicine*, *68*(8), 1472–1480.

Drury, S., Williams, H., Trump, N., Boustred, C., Gosgene, Lench, N., Scott, R. H., & Chitty, L. S. (2015). Exome sequencing for prenatal diagnosis of fetuses with sonographic abnormalities. *Prenatal Diagnosis*, *35*(10), 1010–1017.

Farra, C., Choucair, F., & Awwad, J. (2018). Non-invasive pre-implantation genetic testing of human embryos: An emerging concept. *Human Reproduction*, *33*(12), 2162–2167.

Fung, K. F. K., & Eason, E. (2018). No. 133—Prevention of Rh alloimmunization. *Journal of Obstetrics and Gynaecology Canada*, *40*(1), e1–e10.

Gibbs, N. (2001, February 26). Baby, it's you! And you, and you... *Time Magazine.* http://content.time.com/time/subscriber/article/0,33009,99892,00.html

Gore-Langton, R. E., Nisker, J. A., & Natale, R. (1993). The multipronucleate human pre-embryo: A new model for preimplantation biopsy (Abstract S40). *Society for Gynecologic Investigation.*

Government of Canada. (2004). *Assisted Human Reproduction Act* (S.C. 2004, c. 2). https://laws-lois.justice.gc.ca/eng/acts/A-13.4/page-1.html#h-6052

Handyside, A. H., Kontogianni, E. H., Hardy, K., & Winston, R. M. (1990). Pregnancies from biopsied human preimplantation embryos sexed by Y-specific DNA amplification. *Nature*, 344(6268), 768–770.

Handyside, A. H., Pattinson, J. K., Penketh, R. J., Delhanty, J. D., Winston, R. M., & Tuddenham, E. G. (1989). Biopsy of human preimplantation embryos and sexing by DNA amplification. *Lancet*, 1(8634), 347–349.

Hawking, S. W. (1988). *A brief history of time: From the big bang to black holes*. Bantam Books.

Higgins, M. (2001, March 10). Scientists plan human clones. *National Post*, A1.

Human Cloning Prohibition Act of 2003. (2003). Federal Government of the United States. http://www.gpo.gov/fdsys/pkg/BILLS-108hr534eh/pdf/BILLS-108hr534eh.pdf

Human Fertilisation and Embryology Act 1990. (1990). Government of the United Kingdom. http://www.legislation.gov.uk/ukpga/1990/37/contents

Joseph, M., Saravanabavan, S., & Nisker, J. (2018). Physicians' perceptions of barriers to equal access to reproductive health promotion for women with mobility impairment. *Canadian Journal of Disability Studies*, 7(1), 62–100.

Klemetti, R., Sevon, T., Gissler, M., & Hemminki, E. (2005). Complications of IVF and ovulation induction. *Human Reproduction*, 20(12), 3293–3300.

Lane, A., & Nisker, J. (2016). "Mitochondrial replacement" technologies and human germline nuclear modification. *Journal of Obstetrics and Gynaecology Canada*, 38(8), 731–736.

Langlois, S., Brock, J. A., Wilson, R. D., Audibert, F., Brock, J. A., Carroll, J., Cartier, L., Gagnon, A., Johnson, J. A., Langlois, S., Macdonald, W., Murphy-Kaulbeck, L., Okun, N., Pastuck, M., & Senikas, V. (2013). Current status in non-invasive prenatal detection of Down syndrome, trisomy 18, and trisomy 13 using cell-free DNA in maternal plasma. *Journal of Obstetrics and Gynaecology Canada*, 35(2), 177–181.

Law no. 94-654 of July 29, 1994 Relating to the Donation and Use of Elements and Products of the Human Body, Medically Assisted Procreation and Prenatal Diagnosis. (1994). French Government. *International Digest of Health Legislation*, *45*(4), 473–482.

Levin, I. (1976). *The boys from Brazil*. Random House.

Lo, Y. M., Corbetta, N., Chamberlain, P. F., Rai, V., Sargent, I. L., Redman, C. W., & Wainscoat, J. S. (1997). Presence of fetal DNA in maternal plasma and serum. *Lancet*, *350*(9076), 485–487.

Mullis, K., Faloona, F., Scharf, S., Saiki, R., Horn, G., & Erlich, H. (1986). Specific enzymatic amplification of DNA in vitro: The polymerase chain reaction. *Cold Spring Harbor Symposia on Quantitative Biology*, *51*, 263–273.

Mykitiuk, R., & Nisker, J. (2010). Social determinants of "health" of embryos. In J. Nisker, F. Baylis, I. Karpin, C. McLeod, & R. Mykitiuk (Eds.), *The "healthy" embryo: Social, biomedical, legal and philosophical perspectives* (pp. 116–135). Cambridge University Press.

Newton, C. R., McDermid, A., Tekpetey, F., & Tummon, I. S. (2003). Embryo donation: Attitudes toward donation procedures and factors predicting willingness to donate. *Human Reproduction*, *18*(4), 878–884.

Nisker, J. (2001). Orchids: Not necessarily a gospel. In J. Murray (Ed.), *Mappa mundi: Mapping culture/mapping the world* (pp. 61–110). University of Windsor Press. http://www.uwindsor.ca/hrg/mappa-mundi-mapping-culturemapping-the-world-table-of-contents-0

Nisker, J. (2012). *From Calcedonies to Orchids: Plays promoting humanity in health policy*. Iguana Books.

Nisker, J. (2015). The latest thorn by any other name: Germ-line nuclear transfer in the name of "mitochondrial replacement." *Journal of Obstetrics and Gynaecology Canada*, *37*(9), 829–831.

Nisker, J. (2021). A brief and personal history of "what's in a name" in reproductive genetics. *Medical Humanities*, *47*(2), 228–234.

Nisker, J. A. (1992). Research on human pre-embryos. *ACOG Current Journal Review*, 5(1).

Nisker, J. A., & Gore-Langton, R. E. (1995). Pre-implantation genetic diagnosis: A model of progress and concern. *Journal of Obstetrics and Gynaecology Canada*, 17(3), 247–262.

Nisker, J., Baylis, F., Karpin, I., McLeod, C., & Mykitiuk, R. (Eds.), *The "healthy" embryo: Social, biomedical, legal and philosophical perspectives*. Cambridge University Press.

Practice Committee of Society for Assisted Reproductive Technology; Practice Committee of American Society for Reproductive Medicine. (2008). Preimplantation genetic testing: A Practice Committee opinion. *Fertility and Sterility* 90 (5 Suppl): S136–S143. https://www.fertstert.org/article/S0015-0282(08)03469-9/fulltext

Practice Committee of the American Society for Reproductive Medicine. (2003). Ovarian hyperstimulation syndrome. *Fertility and Sterility*, 80(5), 1309–1314.

Prohibition of Human Cloning for Reprodutction Act 2002. (2002). Australian Government. http://www.comlaw.gov.au/Series/C2004A01081

Randal, A. E. (2004). The personal, interpersonal,and political issues of egg donation. *Journal of Obstetrics and Gynaecology Canada*, 26(12), 1087–1090.

Renga, B. (2018). Non invasive prenatal diagnosis of fetal aneuploidy using cell free fetal DNA. *European Journal of Obstetrics & Gynecology and Reproductive Biology*, 225, 5–8.

Schenker, J. G., & Ezra, Y. (1994). Complications of assisted reproductive techniques. *Fertility and Sterility*, 61(3), 411–422.

Shakespeare, W. (1734). *Romeo and Juliet. By Mr. William Shakespeare*. Eighteenth Century Collections Online. Gale. https://www.gale.com/primary-sources/eighteenth-century-collections-online

Swanton, A., Storey, L., McVeigh, E., & Child, T. (2010). IVF outcome in women with PCOS, PCO and normal ovarian morphology. *European Journal of Obstetrics & Gynecology and Reproductive Biology, 149*(1), 68–71.

Van den Veyver, I. B., Patel, A., Shaw, C. A., Pursley, A. N., Kang, S. H., Simovich, M. J., Ward, P. A., Darilek, S., Johnson, A., Neill, S. E., Bi, W., White, L. D., Eng, C. M., Lupski, J. R., Cheung, S. W., & Beaudet, A. L. (2009). Clinical use of array comparative genomic hybridization (aCGH) for prenatal diagnosis in 300 cases. *Prenatal Diagnosis, 29*(1), 29–39.

Vanstone, M., King, C., deVrijer, B., & Nisker, J. (2014). Non-invasive prenatal testing: Ethics and policy considerations. *Journal of Obstetrics and Gynaecology Canada, 36*(6), 515–526.

Vanstone, M., Yacoub, K., Winsor, S., Giacomini, M., & Nisker, J. (2015). What is "NIPT"? Divergent characterizations of non-invasive prenatal testing strategies. *AJOB Empirical Bioethics, 6*(1): 54–67.

Wright, C., & Burton, H. (2009). The use of cell-free fetal nucleic acids in maternal blood for non-invasive prenatal diagnosis. *Human Reproduction Update, 15*(1), 139–151.

Wright, C., Wei, Y., Higgins, J., & Sagoo, G. (2012). Non-invasive prenatal diagnostic test accuracy for fetal sex using cell-free DNA a review and meta-analysis. *BMC Research Notes, 5*(1), 476.

Chapter 17

Ruth

Calcedonies are rocks
Crusty-surfaced rocks
Rocks that open to onyx and amethyst
Chrysoprase and agate
To become amulets and talismans
Bookends and paperweights
For into each calcedony's core
Millennia poured melted magic

Calcedonies depend on humans
To endow them wisdom or courage
Healing powers or spiritual powers
Ferocity or peace
For each calcedony is unique
Speaks its singular dialect
Seeks its singular respect
As its dignity connects

Friends give Ruth calcedonies
Like the Spanish melon-size bookends
That have bookended her computer
Since her paper books ended
But her computer patiently waits[393]
For Ruth to press its power button
Just as it has waited each day for 17 years
After Ruth turned her last page

Ruth's bookends resemble
Her brain's magnetic resonance image[394]
Complete with fluid-filled ventricles
And crenulated cerebral cortex
With white flecks that attest to the progress
Of the condition disconnecting
Ruth's brain from her voluntary muscles[395]
Except those that open her eyes and move her chin in all directions

Ruth's chin muscles permit her to speak
Albeit quietly and "rarely heard"[396]
And permit Ruth to eat
Albeit slowly and with assistance
And "more important" Ruth's chin muscles wave
The magic wand that propels her
The joystick on her powerchair
Her "chariot of disaster"

---

[393] *Patiently Waiting For…* is the title of the novel Ruth encouraged me to write about her (Nisker, 2015).
[394] MRI
[395] "Voluntary" muscles are the muscles we can control.
[396] The exact words of the woman I call Ruth appear in quotation marks.

Ruth suffers pressure sores
Where her chair sandwiches her skin
Because she can't sense pressure "down there"
And shift her weight off the pressure
Pressure sores are often called bedsores
But Ruth takes umbrage with this word
And assures, "My bed never gets sore"
When the word is uttered by doctors or nurses

When Ruth's pressure sores get infected
She is admitted for intravenous antibiotics
Ruth once asked a nurse to show her
What an infected bedsore looks like
The nurse needed two mirrors for Ruth to observe
Concentric circles of purple and black
And red and yellow "crud"
"Sort of an angry archery target"

When Ruth is with us and recovered enough
From the bedsore-sepsis relentless in sending her here
Each day Ruth patiently waits[397]
To be lifted onto her powerchair
And have her chin velcroed to its joystick
Then Ruth begins haunting the Hospital halls
Hunting doctors at full throttle
Making a blood sport of it

---

[397] *Patiently Waiting For...* is the title of the novel Ruth encouraged me to write about her (Nisker, 2015).

When Ruth spots a doctor she halts her chair
Slowly moves her chin to turn its front wheels
In the direction of the doctor's white coat
As if it is a matador's red cape
She imagines her left foot[398] stamping bullring sand
Then Ruth pushes her chin forward
Charging at the matadoctor
A ferocious bull on wheels

Doctors never notice her charge
Because "patients are invisible to doctors"
But when Ruth is almost on one of us
Terror grips as we plaster the nearest wall
But unlike murderous matadors
Matadoctors never beckon a second charge
Though like matadors kill bulls in the end
Matadoctors killed Ruth in the end

Ruth's "life savings" bought a newer computer
That her chin could function through her chair's joystick
And "the latest in voice-activated software"
To write the poems collected in her heart
But Ruth needs a biomedical engineer
To connect her chin with her computer
And had been languishing on a three-year waiting list
When she drove her chair into my kneecaps

---

[398] I specifically have Ruth imagine she uses "her left foot" in homage to the play *My Left Foot* by Christy Brown, which was made into the Oscar-winning film of the same name directed by Jim Sheridan and starring Daniel Day-Lewis.

The next day I limp through the front door
Of our Hospital's mammoth amphitheatre
Seconds before I am scheduled to begin
Our weekly "Monday Night Narratives" exploration[399]
However I am riveted by Ruth
Whose chair is proximate to the front door
The only door wheelchairs can enter
And then there are steps but no ramps

I feel pressure to begin but am confined by questions
How did Ruth talk a doctor into a "can leave ward" order
Or was no order ever written
And nurses are searching the Hospital for her
How did Ruth learn of "Monday Night Narratives"
From a med student or nurse or a poster
How did Ruth know the amphitheatre's location
Was it from being displayed here over the years

I hear "Doc do you like my slippers?"
And stare down at bear-paw slippers
I hear "Doc I've had these for years"
And fear Ruth is purposefully throwing me off
Of course slippers don't wear out when not walked on
Or was Ruth applauding the nurses for not losing them
I refocus and take a deep breath
"We have a guest with us tonight"

---

[399] "Monday Night Narratives" was a two-hour interactive program I started in 1996 to "imbue compassion in medical students" (Nisker, 1997)

After the medical students leave
Ruth asks me to retrieve her cigarettes
From the sack on the back of her chair
I sigh, "Let's go outside in the fresh air"
Nine feet[400] from the Hospital I reluctantly agree
To place a cigarette in Ruth's lips
And carefully light it with four matches
Before Ruth takes a drag and asks me to write her story

Ruth tells me that a month previously
Sepsis spread through her body from an infected bedsore
And by the time an ambulance brought her to Emerg
She appeared unconscious to the doctors
But Ruth was not unconscious just too exhausted
To open her eyes or respond to questions
And as she couldn't feel them pinching her
She was unresponsive to painful stimuli[401]

Yet Ruth could hear them debate her fate
"Death with dignity or a persistent vegetative state"
A "bed-blocker" in our expensive care unit[402]
Intended for "higher quality of life"[403]

---

[400] Nine feet is the distance to where a person can smoke, as posted on a sign. Ruth stopped her powerchair right in front of the sign.

[401] Pinching and needling skin are part of a standard neurological assessment to determine "response to painful stimuli."

[402] The "Expensive Care Unit" is the term too-often used by physicians in a semi-derogatory manner for the Intensive Care Unit.

[403] "Higher quality of life" became the triage consideration for ICU admission during the COVID-19 pandemic. See Chapter 20, Webinar Physicians' Cavalier Terms for Triage from COVID Ventilators of Persons with Disabilities.

Ruth was concerned Quality of Life Assessments[404]
Could trap persons "like her" in lethal traps
So she considered an Advance Directive
That everything be done for her "no matter what"

But Ruth heard of danger with Advance Directives
And worried written words could be used against her
To foster her death with dignity
So she firmly reconsidered
Then Ruth heard of Life Story Decision Making
For which you carry a card with contact information
Of persons you trust to make decisions for you
Ruth calls her card her "life preserver"

I come home from a distant conference
To see my phone's red light flashing
And hear my answering machine tremble
"Ruth's in trouble, come to the Hospital"
I assume the woman who called
Is on Ruth's list of trusted persons
Trusted to insist all be done for her
As Ruth wants "to live no matter what"

---

[404] "Quality of Life Assessments," lightly called "qualies" by some Emerg and ICU physicians, residents and even medical students, predict the worth of living a disabled condition through the eyes of "able-bodied" assessors on tick box and short answer paper.

I drive urgently to the Hospital
Leave my car at the ER doors
Dash in and look at "The Board"[405]
Ruth's name is not there
I ask to which floor she's been admitted
But their computer is taking too long
So I run up the stairs to Intensive Care
Worrying and hoping Ruth is there

The nurse at the desk seems to expect me
Her left index finger is pointing
To a draped-off bed at the Unit's far end
Where incandescent curtains project ominous shadows
I turn to run to the curtains
But the nurse grips my left wrist
And urges "Please stay with me
At least till they're finished"

I extricate my wrist and run
And fling open the curtains … to horror
Bacteria has necrosed Ruth's skin
And swollen her body to a huge balloon
Knotted at her neck and elbows and wrists
By a sadistic birthday-party clown
Ruth's closed eyes bulge black tennis balls
Ruth's chin is gone

---

[405] "The Board" occupied a wall with the names of persons "patiently waiting" (Nisker, 2015) in the ER during various stages, including "waiting" to be seen by a physician, "waiting" to have blood drawn, "waiting" to receive results of their blood tests, "waiting" to be sent to radiology, "waiting" to receive x-ray results, "waiting" to be referred to a specialist, "waiting" to be admitted, "waiting" to be sent home.

The periphery of my underwater vision
Sees an ICU doc on my left
Acknowledging me as he draws
Lethal drugs into syringes
Hollowness expands within me
Vacuuming me downward
I fight the hollowness with I can stop this
Ruth wants "to live no matter what"

I focus hard at Ruth
Working hard to see
Working hard to breathe
Drowning in what I see
I notice a blurred woman sitting to my right
As I turn left to the ICU doctor
"You can't disconnect her
Ruth wants to 'live no matter what'"[406]

He puts his hand on my shoulder
Whispers "There's nothing left of her"
I retort "How can you be sure
Ruth has appeared dead before"
He turns my shoulders "Jeff look at her
There's nothing left of her"
I plead again "How can you be sure"
But he just goes back to his syringes

---

[406] See Chapter 20: Webinar Physicians' Cavalier Terms for COVID-Ventilator Triage of Disabled Persons

I put my body between Ruth's and his
As Ruth's Life Story Decision Making suggests I should
But instead of insisting Ruth remain on the ventilator
I place my lips where Ruth's right ear should have been
I plead "Ruth give me a sign
Twitch an eyelid or move your chin
Try to move something
Help me Ruth please"

Earlier versions of Ruth's story were published as the short story "Chalcedonies" in the *Canadian Medical Association Journal*,[407] and in *From the Other Side of the Fence: Stories from Health Care Professionals*.[408] I then expanded Ruth's story to a full-length play that was produced in several countries, and published in my compilation *From Calcedonies to Orchids*,[409] as well as chapters in *Reflective Practice*[410] and *Health and Humanities Reader*.[411] I was able to honour Ruth better with the novel *Patiently Waiting For....*"[412]

References

Nisker J. Calcedonies: critical reflections on writing plays to engage citizens in health and social policy development. Reflective Practice, 2010 Sep; 11 (4): 417-432

---

[407] Nisker, 2001.
[408] Nisker, 2008.
[409] Nisker, 2012.
[410] Nisker, 2010.
[411] Nisker, 2014.
[412] Nisker, 2015.

Nisker J. Calcedonies (chapter 42). In: Jones T, Wear D, Friedman LD, editor(s). Health and Humanities Reader. (United States): Rutgers University Press; 2014.

Nisker J. Chalcedonies. In: Nisker J. editor(s). From the Other Side of the Fence: Stories from Health Care Professionals. Halifax (Canada): Pottersfield Press; 2008. p.172-176

Nisker, J. (2012). *From Calcedonies to Orchids: Plays promoting humanity in health policy.* Iguana Books.

Nisker, J. (2015). *Patiently Waiting For...* Iguana Books.

Chapter 18

Victor

Victor was a gentle giant, the strongest and kindest man I ever met. Victor's physical strength was genetic, augmented by growing up on a farm and years of lugging a welding tank. Victor's kindness was non-genetic, but an integral part of his nature. Victor's strength combined with his kindness to make him the go-to person for his many siblings and friends when they needed assistance, be it building a shed in their yard or brightening their spirits with hospital-bed visits. Even after Victor's years of welding collapsed discs in his lower back, making walking excruciating, he fought through the pain, never complaining, always there for others. Yet I never appreciated how strong Victor was until he took on pancreatic cancer.

    Victor did not complain of the new pain growing in his abdomen. It was not until family members observed Victor's skin had turned yellow that he agreed to see a doctor. Two months later, a CAT scan concluded it was pancreatic cancer. Victor was referred to a surgeon, who thought the location of Victor's tumour, blocking his bile duct and thus causing jaundice, might have declared his pancreatic cancer early enough for cure; a possibility almost unheard of for pancreatic cancer. With this possibility of cure, Victor endured a "Whipple procedure," perhaps the most onerous of cancer surgeries.

The planned four-hour procedure took more than eight hours. The surgeon, on leaving the OR, reported to Victor's family that Victor's cancer was more aggressive than anticipated. The surgeon's word "aggressive," reported to me long-distance by his daughter Roxanne, echoed another surgeon's word "aggressive," reported to me outside another OR after my Mother's mastectomy.[413] Victor's surgeon's "aggressive" reverberated with a similar death knell.[414]

Victor's strength permitted his rapid recovery from the massive surgery, and he soon resumed his daily rounds of assisting family members and friends. Three months after his surgery, Victor was summarily told by the surgeon[415] that the cancer had recurred. Victor's strength soon endured the wrath of pancreatic cancer's advance through his abdomen, invading his liver and bowel. I knew I would never possess Victor's strength.[416]

Victor was the second youngest of eight children in a Ukrainian family. His parents immigrated to a small farming community in the Canadian prairies in 1925, where Victor was born a decade later. Victor eventually had to leave the family farm to try to find employment in nearby Edmonton, which he succeeded in doing as a pipe welder. Victor was always eager to go to work before dawn, even on frigid Edmonton mornings, and more eager to return to his wife and two daughters, with whom his quietness basked. The lunch box that Victor took to work sits proudly beside his picture on Roxanne's desk. Victor remained rooted in the family farm, returning there frequently on weekends, until metastatic pancreatic cancer made returning impossible. Victor's family's homestead still exists, but so

---

[413] See Chapter 10, She Lived with the Knowledge; Nisker, 2012.

[414] See Chapter 10, She Lived with the Knowledge.

[415] My mother's surgeon similarly could not communicate compassion. It was similar for the first surgeon I worked with as a medical student, who asked me to tell one of his patients for him that lung cancer had recurred, and death was imminent.

[416] See Chapter 25, The Arrogance of "But All You Need Is a Good Index Finger."

does the phantom presence of his family's pancreatic-cancer gene; the gene responsible for killing Victor and three of his siblings.

Like with many immigrant families, most of the upper branches of Victor's family tree have been severed.[417] When two of Victor's older siblings died from pancreatic cancer like their father, there was a suppressed suspicion that pancreatic cancer ran in Victor's family. When Victor's younger sister received her pancreatic cancer diagnosis, it became clear, at least to me, that Victor's family carried a rare pancreatic cancer gene and that this gene was "autosomal dominant" and of "high penetrance."[418] The gene being autosomal dominant and of high penetrance, combined with the deadliness of pancreatic cancer, dooms half of the next generation of Victor's large family to die from pancreatic cancer.[419]

Victor was proud of his Ukrainian heritage, as were all his family members. Roxanne speaks with great love of her "*Baba*," whom Roxanne looked forward to being with every weekend of her childhood. Roxanne's love of her *Baba* reminisces my love of my *Bubie*;[420] women from the same part of the Ukraine who came to Canada at approximately the same time. Roxanne's *Baba* and my *Bubie* cooked similar "old-country" foods like *halipses*, *latkes*, *blintzes*, and *borscht*; but *perogies* were her *Baba*'s only.

Victor generously gave me the gift of his kindness, even when I was bothering him on an Algonquin dock while he was trying to quietly fish. The vision of Victor at the end of the dock, fishing rod extended over the lake, patiently waiting in blissful peace, permeates me with

---

[417] See Chapter 6, Miriam; See Chapter 10, She Lived with the Knowledge.

[418] "Autosomal dominant" means you only need to inherit one of the two copies of a gene to potentially develop the related condition; "high penetrance" means that if you inherit the gene, you are very likely to develop the condition. Only 5 percent of pancreatic cancers are hereditary, and less than half of the causative genes have been discovered.

[419] See Chapter 15, The Injustice of Needing Angelina Jolie.

[420] See Chapter 2, "You Must Go to Medical School or Hitler Will Have Won"; See Chapter 10, She Lived with the Knowledge; See Chapter 13, The "Helix of Life" Revisited: DNA in Concrete and Not.

warmth to this day. Victor's caring smile never diminished with my questions, and always broadened when he brought in a smallmouth bass. Victor's smile broadened further each time I asked him about hockey's famous "Uke line," consisting of his heroes Vic Stasiuk, Johnny Bucyk, and Bronco Horvath. Victor is my hero.

I have a picture of Victor holding two large fish on a dock amidst a coniferous forest. His picture sits on a dresser beside my Mother's picture. On the other side of Victor's picture is a quadtych of Roxanne, taken by her sister on the Thanksgiving following Victor's surgery. Victor knew this would be his last Thanksgiving and was giving thanks to his wife and daughters by taking them to Jasper National Park. In all pictures of the quadtych, Roxanne's face is in profile, almost kissing the grey-whiskered dandelion she is holding. In the first picture, Roxanne is inhaling. In the second picture, Roxanne is beginning to blow at the dandelion's whiskers. In the third picture, the dandelion's stem bends in the breeze of her breath. In the final picture, Roxanne's lips purse in a mix of serene smile and trembling concern as she is about to kiss the few remaining whiskers of the dandelion. Roxanne was making a wish; a wish for Vic.

I had the gift of being able to make it to Victor's hospital bed the evening before he died. I carefully took Victor's intravenoused right hand in mine, then received another great gift from Victor, as he slowly opened his eyes and embraced me in the warmth of his smile. Somehow Victor had the strength to acknowledge my presence through his pain and sedation, and the kindness to convey he was glad I had made it to Edmonton to say goodbye to him, amidst my own cancer treatment.[421]

Roxanne and I remained with Victor as long as we were permitted; Roxanne gently hugging her father's neck with her head on his chest. We then checked into a nearby motel, where we held each other's trembling bodies beneath a blanket. Roxanne was

---

[421] See Chapter 25, The Arrogance of "But All You Need Is a Good Index Finger."

worrying she would never touch her father alive again. I was also worrying that Roxanne would never touch Victor alive again; but I confess that I was also worrying about the impact of my oncologist's verdict earlier in the day, a four-hour plane flight away, on my prostate biopsies that were reported as "aggressive" cancer, and indicated I likely possessed a BRCA gene mutation.[422] Thus, in addition to worrying about Roxanne and Victor, I was worrying that I had passed the BRCA gene mutation on to at least one of my sons, and, more importantly, to more than one of my granddaughters. Roxanne and I absorbed each other's quivers until four in the morning, when the phone rang. Roxanne quickly picked up the receiver. Victor was dead. We put on our coats and boots, and headed back to the hospital to touch Victor one last time.

I could not remain in Edmonton for Victor's memorial service eight days later. Not being there is one of the regrets of my life, but I had promised my radiation-oncologist, who did not want me to go to Edmonton and delay treatment, that I would return as soon as possible.[423] I regret not being there for Roxanne, not being there for her mother and sister, not being there for Victor. I had made an error, and, for the rest of my life, will regret not having been there.

The day after I returned home, I had an MRI to make sure there were no metastases invisible to the CAT scan–planned radiation field. Only because I was a physician working in a cancer hospital was an MRI available to me. I received the results of the MRI on the day of Victor's memorial service, indicating a thumb-sized metastasis outside of the planned radiation field. This metastasis dramatically worsened my prognosis. I called Roxanne later that day to make sure she was okay after Victor's memorial service. She was. Roxanne is

---

[422] A BRCA gene mutation caused the early onset breast cancer that killed my Mother and Grandmother; See Chapter 6, Miriam; See Chapter 10, She Lived with the Knowledge; See Chapter 13, The "Helix of Life" Revisited: DNA in Concrete and Not.

[423] See Chapter 25, The Arrogance of "But All You Need Is a Good Index Finger."

strong. I debated whether to share my MRI's results, and only because Roxanne is strong did I tell her about the large metastasis. At that moment, oxygen was vacuumed from my being and replaced with the fear that I would not be able to handle cancer's ravage nearly as well as Victor did. I shared this overwhelming fear with Roxanne, whose silence confirmed my fear's validity. I broke the silence with, "I'm not as strong as Vic, I wish I was, but I know I'm not." Roxanne did not try to disabuse me of this reality. I took some comfort in adding, "I don't know anyone who is as strong as Vic."

I remember with fondness and sadness the year after Victor's death, when Roxanne gave me a shopping list to be filled at a Ukrainian church near the hospital where I work. Roxanne had decided to have Christmas in Ontario for the first time, rather than at her parents' home in Edmonton. She had found this Ukrainian church online, and provided me the address and the two-hour window in which orders could be picked up the afternoon before Christmas.[424] As the church came into view, I was amazed by its massive majesty, and even more amazed that such majesty existed less than a kilometre from where I had practised for over 30 years. This was not a Ukrainian church; it was an enormous Ukrainian cathedral, complete with a magnificent dome.

As I drove up the circular driveway, I saw "pick-ups" arrows pointing to the cathedral's side door. Opening that door, I saw more arrows directing me downstairs, where I opened a door to a huge banquet hall filled with long tables and perimetered with kitchens and dishwashing rooms. I picked up the many boxes of Ukrainian food that Roxanne had ordered, taking several slow trips to my car to

---

[424] I had not previously heard of the street where the church was located, but assumed it was the church under the two onion domes that were visible from my office window. It was not. This church turned out to be a Russian church, and was locked. I forgot the address in my office, and it took asking several passersby before one knew where the Ukrainian church was located. Several Ukrainian churches later, I was directed to a Ukrainian church very close to the hospital, but in the opposite direction from which I was searching.

gently place them inside. After all the boxes were loaded in, I walked back through the cathedral's side door, but this time went up the stairs and opened a door to the cathedral's main sanctuary. I knelt in the pew nearest the door, basking in stained-glass sun, basking in Victor. I expressed gratitude for the gift of knowing Victor, of knowing Victor's kindness, of knowing Victor's strength. As I drove home, I was overwhelmed with the essence of Victor. The windshield seemed blurred by rain.

Roxanne, her mother, sister, brother-in-law, and I gathered in the kitchen the next day. Roxanne had organized everything in advance, and now assigned each of us a task. Roxanne already had a turkey in the oven and other recognizable Christmas foods stationed in various places in the kitchen. Roxanne had also warmed the Ukrainian delicacies from the cathedral, and had them prominently displayed on the dining room table. There were *perogies, halipses, blintzes*, and *latkes*, as well as Ukrainian foods of which I had no previous knowledge. This meal would be a tribute to Victor. He would be toasted with every glass of wine and water, and celebrated in every hour of Christmas-dinner conversation.

As I am writing Victor's story five years after his death, one of Victor's nieces is dying of pancreatic cancer. She is exactly Roxanne's age, and I am worried about Roxanne. As I am writing Victor's story, I am upset about the injustice of the reluctance to fund research on hereditary pancreatic cancer; an injustice permitted because hereditary pancreatic cancer is a rare form of pancreatic cancer, and pancreatic cancer is a rare form of cancer, though the second most common of all cancer-killers. I believe this injustice is permitted because of the supposed cost-effectiveness of cancer research-funding agencies, and the definite corporate-profit effectiveness of biotech companies, both of which have abandoned individuals and families to the horrors of pancreatic cancer.[425] Roxanne has a significant

---

[425] The lack of research to determine the rare hereditary pancreatic cancer genes stands in sharp contrast to the quantity of research that determined the much more common hereditary breast cancer (BRCA) genes, resulting

chance of developing pancreatic cancer, as do her sister and her many first cousins. Roxanne is taking a leadership role in encouraging her family members to begin surveillance-imaging[426] and to have genetic counselling.

Victor, the strongest man I have ever met, was still gentle enough to create with his huge hands the tiny and fragile twig sculpture that graces our home. This sculpture reminisces a tree bare of leaves standing up to a strong wind. Though just a few centimetres in height, Victor's sculpture will always be a giant presence within our home, and within Roxanne, and within me.

References

Nisker, J. (2012). *From Calcedonies to Orchids: Plays promoting humanity in health policy.* Iguana Books.

---

in biotech companies such as Myriad Genetics making billions. Victor's family should be viewed as similar to BRCA gene families, but with the understanding that current testing capabilities for pancreatic cancer genes reveal only half of the genes that exist in high-risk families, considering the autosomal-dominant inheritance pattern.

[426] Magnetic resonance cholangiopancreatography (MRCP) and esophageal ultrasound exist in Canada, but are generally either unavailable or associated with long waiting lists.

Chapter 19

Canadian COVID Injustice on Beaches and Beach Volleyball Courts

In January 2022, the Canadian Broadcasting Corporation reported unmasked "twenty-somethings" revelling in the aisle of a Sunwing flight from locked-down Montreal to Cancun's wide-open beaches.[427] Their drinking and vaping, crowd-surfing and cheering, were celebrating the two-week COVID-induced delay in resumption of classes after Christmas break to soon break out towels on Mexican beaches, even though Mexico remained one of the world's severest COVID hot spots.[428]

The students seemed to have no regard for public health warnings that travellers would be bringing back to Canada the highly transmissible Omicron variant[429] they were air-bound to catch in Mexico's tightly-packed bars, restaurants, dance clubs, and gyms, not

---

[427] Marchand, 2022.
[428] Mendoza, 2022.
[429] The Omicron variant was the most contagious strain of COVID-19, and became the dominant strain by the beginning of 2022 (Khandia, 2022).

to mention proximate-towelled beaches, and spread on the tightly-packed jets returning them to Canada. Canadian air carriers also had no regard for public health warnings; rather were ecstatic their jets were permitted to take Canadians to Mexican beaches, and of course back, even though some were bound to transport back the souvenir of COVID. Like many Canadian physicians in January 2022, I had assumed the air above our closed-to-car border was as humidor-sealed as our land border, and was abhorred by press-pictures premoniscing our looming public-health disaster.[430]

The twenty-somethings on this Sunwing flight should not be singled out for condemnation, as during the pandemic many Canadian beach-seekers ignored public health proclamations, including those captured on national television in Calgary's airport wearing suntans and sombreros as they transferred off their flight from Mexico onto domestic flights.[431] This lack of Canadian community consciousness was also prominent on Canada's beaches during our two COVID summers, as social distancing, though prescribed, was not enforced. Lack of community consciousness would continue into our third COVID summer, in which "COVID fatigue" would make us believe that it was over.

This prescription for social distancing reminisced the social distancing my Grandmother[432] did enforce on Toronto's Sunnyside Beach when I was a child because of a much less contagious virus, polio. My Grandmother was quick to shoo away children who wanted to play with me before they came within six feet.[433] She even insisted

---

[430] Marchand, 2022.
[431] Canadian Broadcasting Corporation News, January 2022.
[432] My Grandmother was a second mother to me. She died from hereditary early onset breast cancer when I was a teen. I revisit hereditary breast cancer in Chapter 2, "You Must Go to Medical School or Hitler Will Have Won"; Chapter 5, Princess Margaret; Chapter 6, Miriam; Chapter 10, She Lived with the Knowledge; Chapter 13, The "Helix of Life" Revisited: DNA in Concrete and Not; and Chapter 15, The Injustice of Needing Angelina Jolie.
[433] Two meters or just more than six feet was the distance Canada's Public Health Agency declared as likely safe from inhaling COVID when wearing a

I pee in Lake Ontario rather than in the bathrooms in Sunnyside Pavilion, which of course would have been fine with me had she not stood guard beside me. My Grandmother stood guard in front of me after she sat me firmly at the back of almost-empty TTC[434] streetcars from the streetcar barn[435] to Sunnyside Beach and back.

Although the polio virus is much less contagious than the COVID-19 virus, particularly its Omicron subvariants,[436] it tends to have more serious consequences, particularly paralysis, for children.[437] There were other viruses just as contagious as COVID prevalent when I was a child, including measles and chicken pox; however, these viruses had such limited consequences for children that we were sent out to play with friends with spots so that as adults we would not suffer their more serious consequences.

One autumn, the children in our school were lined up in the gym to receive a needle in our arm called the "Salk Vaccine." We were not thrilled with the prospect of being needled, especially after classmates ahead in the line started screaming even before they were jabbed.

---

mask (Government of Canada, updated Feb 15, 2022). In the prescient 2011 film *Contagion*, directed by Steven Soderbergh, 10 feet was the distance prescribed.

[434] Toronto Transit Commission

[435] The TTC "streetcar barns" are south of St. Clair Avenue West and east of Christie St., near where my Grandmother's tiny house was located. My Grandmother once arranged a tour of the massive structures, filled not only with streetcars, but with large maintenance and repair equipment, including huge cranes. The streetcar barns were at the same time magical and overwhelming for a young child.

[436] Omicron subvariants are the most contagious strains of COVID-19 (Khandia, 2022; Pelley, 2022).

[437] The paralysis caused by the polio virus was poignantly brought to 21st century audiences by the 2017 biographical film *Breathe*, directed by Andy Serkis, with Andrew Garfield portraying Robin Cavendish, a young man who contracted polio. One of the scenes overwhelms the audience with the injustice of warehousing persons with polio, stacked in layers of iron lungs with only mirrors to allow them glimpses of a small fraction of the warehouse floor.

Jonas Salk was a physician whose vaccine was introduced in schools in the United States in 1955, and over the next few years in Canada.[438] The Salk Vaccine was replaced in 1962 by the "Sabin Vaccine,"[439] which we kids thought was great because it was a sweet pink drink out of a tiny white paper cup rather than the booster needle we had been enduring annually to ensure herd immunity.[440] However, it was Dr. Salk who continued to be lauded to the point of legend. Dr. Salk was my childhood hero.

Even after I received Dr. Salk's vaccine, my parents continued their concern I would catch polio, as Dr. Salk's vaccine, like the COVID-19 vaccine, did not come with a 100-percent written guarantee like our Westinghouse refrigerator. My parents concern led to avoidance of Sunnyside Beach, streetcars, and crowded city streets. When I was eight they packed me north to the outdoor space of a children's camp, where my parents perceived "fresh air" would lessen the chance of catching the polio virus. My parents were likely unaware we would be bunking indoors in close quarters rather than breathing the virus-free idyllic air of forests and fields. Bunking indoors in close quarters led to COVID becoming endemic in Canada's long-term care facilities.[441] Bunking indoors in close quarters led to COVID growing rampant in the barracks of Canada's migrant agricultural workers,[442] most of whom did not have access to vaccines. Butchering meat in close quarters led to COVID becoming endemic in Canada's meat-packing plants,[443] including the Cargill

---

[438] By June 1956, Connaught Laboratories in Toronto had manufactured 2.3 million doses of Salk vaccine (Rutty, 2005, p. I-14); Baicus, 2012; See Chapter 13, The "Helix of Life" Revisited: DNA in Concrete and Not.

[439] Baicus, 2012. "The Sabin Vaccine" is no longer recommended in Canada because most cases of polio from 1980 to 1995 were associated with oral polio vaccine, but it continues to be widely used internationally (Government of Canada, 2018).

[440] "Herd immunity" (Rashid, Khandaker, & Booy, 2012).

[441] Phillips, 2021; Webster, 2021.

[442] Ansari, 2020.

[443] Neustaeter, 2020.

plant 20 minutes from my home in which 82 workers became ill with COVID;[444] there were even larger outbreaks in meat-packing plants in Alberta.[445]

At the summer camp, I became friends with a boy who wore metal braces on his legs and used metal crutches that had black-rubber handles protruding out in front. My friend was quick on his crutches across the rocky terrain, and he played tetherball by supporting himself on one crutch while hammering the ball around the pole with the other. It was he who first told me about polio, and that he had needed an "iron lung" to breathe for two years. He described an iron lung as looking like the large propane tank lying flat behind the showers, except that an iron lung has an opening for your face.[446]

I did not think about polio again until I was a medical student spending Friday afternoons in a decrepit chronic care "facility" on the outskirts of Toronto, where I observed persons my age who had been infected by the polio virus using chin-operated wheelchairs,[447] with a few requiring ventilators.[448] It was also as a medical student that I learned that Dr. Salk had refused to patent the polio vaccine to financially profit from stopping the polio virus. When questioned by a journalist as to who owned the patent,[449] Dr. Salk responded, "The people, I would say. There is no patent. Could you patent the sun?"[450] Dr. Salk became my medical school hero.

The COVID social distancing that was prescribed but not enforced on Canada's crowded beaches, nor on the crowded

---

[444] Dubinski & Rodriguez, 2021; Graham, 2021a.
[445] Guse, 2022.
[446] As noted previously, iron lungs were poignantly brought to 21st century audiences by the biographical film *Breathe* in 2017, directed by Andy Serkis (2017), with Andrew Garfield portraying Robin Cavendish.
[447] See Chapter 17, Ruth.
[448] See Chapter 20, Webinar Physicians' Cavalier Terms for COVID-Ventilator Triage of Disabled Persons.
[449] The journalist was Edward R. Murrow, one of the first investigative reporters.
[450] Bos, 2013.

summer streets of large-city "hotspots" like Toronto, brought COVID to our university town within a few weeks of our students coming to campus for the 2020 fall semester.[451] COVID outbreaks quickly swept through the close quarters of our University's residences, often linked to the many lack-of-distancing and lack-of-masking parties, as well as too many sports activities.[452] Our University responded by closing all on-campus bars and sports facilities, and insisting on masking on campus. However, the ominous unmasked, social-distance-lacking lineups to the numerous just-off-campus bars and volleyball beaches forecast the "spike"[453] in COVID on our isotherm map that "slammed"[454] our university town after the first COVID Thanksgiving, with our students returning from large cities. The spike was then sharpened by the many tightly-packed indoor Halloween parties,[455] where masks were much more likely to be worn over eyes than the prescribed mouths and noses. Few tickets were inflicted until a $10,000 fine[456] was delivered along with the pizza to the "host"[457] of a large indoor-unmasked party in a small house just off-campus. The following weekend, the 19-year-old hosts of an even larger indoor-unmasked house party were formally charged, and our region's Medical Officer of Health warned that these parties could have catastrophic public health implications.[458]

---

[451] Dubinski, 2020.

[452] Dubinski, 2020.

[453] "Spike" is the volleyball term for hitting the ball sharply down over the net, ironic for the sharp increase in COVID.

[454] "Slam" is a volleyball term for hitting the ball hard.

[455] Like the party at which 150 unmasked students squeezed into a tiny house (Dubinski, 2020).

[456] Bicknell, 2020, Nov 10.

[457] The term "host" not only acknowledges the person in whose home the party occurred, but also acknowledges the persons whose bodies serve as a home for COVID virus to be growing, and eventually spreading from their lungs to outside through the open windows of their mouths.

[458] Dubinski, 2020.

Yet our students continued to protest infliction of masks, citing "personal freedom"[459] and that this was their only time for their "university experience."[460] Some students also refused COVID testing and contact tracing, again in the name of "personal freedom," while other students refused because of their concern regarding the "stigma" of a positive test.[461] One such student was an editor of our University's student newspaper, who eloquently expressed her fear of "stigma," as well as the disempowerment of women students' ability to protect themselves.[462] Another student spoke of her dilemma when her boyfriend visited from Toronto, and eventually infected her, and through her all her housemates.[463]

The cooler weather in the approaching first Canadian COVID winter encouraged students to move indoors and cluster[464] in the concentrated virus-contaminated air of cafeterias, common rooms, and indoor house parties. Our students also eagerly breathed the virus-contaminated air in off-campus bars, gyms, restaurants, and movie theatres, as well as the malls that our province's premier refused to lockdown in our town pre-Christmas, but had locked down in Toronto and its suburban regions. The Toronto lockdowns brought Christmas-shopping caravans of cars and convoys of chartered buses to the malls of our university town,[465] where the lack of social distancing became dense and intense with this "financial

---

[459] "Freedom" became the mantra for "antivaxxers" (Hotez, 2021; Kurjata, 2021; Lee-Shanok, 2021). In January 2022, antivaxxers represented over 70 percent of persons in our ICUs (Daigle, 2022; Favaro & Jones, 2022), and continued to be an open invitation to new COVID variants. "Freedom" was also the mantra for the antivaxxer-truckers who convoyed across Canada to Parliament Hill in Ottawa, and blocked border crossings to the United States, to protest Canada's vaccine mandate (Andrews & Anand, 2022).
[460] Chatham, 2020.
[461] Hristova, 2020; Rivers, 2020; Paling, 2021.
[462] Dubinski, 2020.
[463] Dubinski, 2020.
[464] Clusters of COVID would increase indoors.
[465] Rocca, 2020.

gift."[466] There was rarely a mask observed. The malls in our university town were shut down on Boxing Day.[467]

Just before our first COVID Christmas, our Public Health Unit warned of the risks of our students taking COVID home to their families from their pre-Christmas "killer parties."[468] At risk were the lives of their grandparents and other elderly or otherwise sub-immune relations, who are not invulnerable like our students see themselves, and indeed are to a large degree regarding death from COVID. When our students returned home for COVID Christmas, their grandparents had no choice but to be willing to risk COVID, as they were so eager to be with their grandchildren at family gatherings.

COVID continued in 2021, with the virus proliferating in our region through more indoor, unmasked house parties, including many Super Bowl parties.[469] Later that February, Reading Week took our students on their traditional party-pilgrimage to Daytona Beach and other Florida beaches, press-picture crowded with proximate towels of well-oiled sun tanners, interrupted intermittently with beach volleyball nets. There seemed to be no consideration of social distancing, but significant consideration of the pale shadow a mask would cast on a beach-tanned face. Canada's airlines were of course overjoyed to still be able to fly over our closed-to-car borders to COVID "hot spots" and back, not concerned with bringing COVID back to Canada. When flights into Canada were terminated at the end of February 2021, Canadian sun-worshippers landed in American border cities[470] and walked across our closed-to-car and closed-to-airplane border to COVID-created Canadian limo services.

---

[466] Rocca, 2020.
[467] Rocca, 2020.
[468] Bogart, 2020.
[469] "Homeowner Slapped with Hefty Ticket as Police Sack Super Bowl Bash," 2021.
[470] These airline passengers landed in Buffalo, Detroit, Vermont, North Dakota, and Seattle in what would have been just a short drive across the now closed-to-car but permeable-to-footstep border to Toronto, Windsor, Montreal, Winnipeg, and Vancouver.

In early March 2021, restrictions were loosened, even in Toronto and other COVID hot spots, for the purpose of "keeping businesses economically sound."[471] This loosening of restrictions started killing persons just as restrictions were proven to be working.[472] Loosening of restrictions overflowed our ICU[473] and spread COVID from floor to floor in our Hospital, at the same time as Canadian news footage observed lack-of-distancing-lack-of-masking post-secondary students[474] on "freedom marches."[475]

I volunteered to work on our Hospital's newly-created COVID-overflow floor, but my requests were resisted by, "We need nurses not doctors." I volunteered to give COVID vaccines at the Western Fair Agriplex, but was told, "There's no vaccine left, and there's none coming till late Spring." I did not volunteer to be one of the physicians making triage plans for when we ran short of COVID-ventilators.[476]

Although many of our students were aware to "Beware the Ides of March,"[477] and were aware to beware gathering two days later on St. Paddy's Day, our Public Health Officer, Mayor, and University

---

[471] Office of the Premier, 2021

[472] Stone & Moore, 2021

[473] In early April 2021, our ICU began receiving patients from outside our regions as the ICUs in Toronto reached over-capacity. Our hospital increased ICU bed capacity on April 16 (Bicknell, 2021; Sharkey, 2021).

[474] "Post-secondary student" became the term universities preferred over "university students" to ensure universities would not be seen as a COVID-breeding ground and diminish fundraising campaigns.

[475] Lee-Shanok, 2021. I take umbrage with the term "freedom marches," as it demeans the freedom marches of the civil rights movement in the American South from 1963 to 1968, as well as demeaning Dr. Martin Luther King Jr. At Dr. King's inspired March on Washington in 1963, he delivered his famous "I Have a Dream" speech, a poster of which inspired from the wall between my children's bedrooms, and does so now from my living room. Dr. King was shot and killed on the eve of his participation in the nonviolent sanitation workers' protest in Memphis Tennessee in 1968.

[476] See Chapter 20, Webinar Physicians' Cavalier Terms for COVID-Ventilator Triage of Disabled Persons.

[477] The Ides of March is March 15, the day Shakespeare's soothsayer warned Julius Caesar to beware of.

President all still felt it important to warn our students that congregating on St. Patrick's Day would create a "super-spreader event."[478] Notwithstanding, emerald waves of unmasked, tightly-packed students crashed over our volleyball beaches to flood our streets with shamrock vests, leprechaun hats, and other emerald-green regalia. The flood spread to the off-campus bars that our premier had just reopened so as not to miss the business of St. Paddy's Day.[479] Subsequently, COVID again spiked occupancy in our ICU,[480] with most cases linked to student partying on St. Patrick's Day. However, the persons admitted to our ICU were not the student partiers; rather cleaners, food servers, and other persons of socio-economic disadvantage working in our University community, who were not in a position to turn down employment.

The warming weather of spring 2021 again saturated our volleyball beaches with unmasked students, whose spike-induced grunts, and post-spike laughter spread COVID in the face of public health warnings that mass contagion would encourage new variants.[481] I often run by one of these volleyball beaches, where teams of six students play each other in intimate, unmasked proximity, while other unmasked teams wait shoulder-to-shoulder on the sidelines. When a spike bounced a fluorescent-green volleyball to my running shoes, one of the students politely asked if I would throw the ball back, as had been my practice prior to COVID. I softly kicked the ball back instead, politely claiming COVID-precaution, and not-so-politely receiving a "whatever" shrug amidst unanimous headshakes. I continued my run past tightly-packed unmasked students basking on beach blankets, and could not help but ponder why our University failed to include credit courses on community consciousness in all

---

[478] Graham, 2021.
[479] Office of the Premier, 2021.
[480] In early April 2021, London's ICUs began receiving patients from outside of London as ICUs in Toronto reached over-capacity, causing our Hospital to increase ICU bed capacity on April 16 (Bicknell, 2021; Sharkey, 2021).
[481] Denette, 2021.

curricula. Of course, our students are but a microcosm of the many self-centred Canadians who privilege "personal freedom"[482] over community responsibility.

On April Fool's Day 2021, our University's president gave up on trying to encourage community consciousness, and sent our students home.[483] On April 8, Public Health finally persuaded our premier to lock down our province, in spite of strong push-back from business. On Canada Day, July 1, 2021, though COVID vaccination of our population was still a work in progress, Canada's beaches again swarmed with unmasked young persons, many of whom were transmitting COVID unknowingly because they did not show severe symptoms of illness. This COVID camouflage "set"[484] another spike of the virus when our students returned in September. During Frosh Week,[485] rivers of students wearing coloured T-shirts representing their faculties or residences marched through campus and surrounding streets, reminiscent of the streams of emerald-green on St. Paddy's Day. As autumn continued, some students added refusal of vaccination to their list of refusals of infringements on their "personal freedom," voicing concerns analogous to other protesters of Canada's vaccine-mandate, who say they are pro-freedom not antivax.[486] However, our province's vaccine mandate

---

[482] "Freedom" became the mantra for the "antivaxxers" (Hotez, 2021; Kurjata, 2021; Lee-Shanok, 2021) who represented over 70 percent of persons in our ICUs in January 2022 (Daigle, 2022; Favaro & Jones, 2022), and who continue to be an open invitation to new COVID variants. "Freedom" also became the mantra for the antivaxxer truckers who convoyed across Canada to Parliament Hill in Ottawa, and blocked border crossings to the United States to protest Canada's vaccine mandate, (Andrews & Anand, 2022); See Chapter 24, Antivaxxer Xenophobic COVID Violence.
[483] Butler, 2021.
[484] "Set" is a volleyball term for a soft high pass leading to a "spike."
[485] "Frosh Week" is formally referred to as "Orientation Week" by most universities.
[486] Lee-Shanok, 2021; See Chapter 24, Antivaxxer Xenophobic COVID Violence.

meant unvaccinated students were barred from campuses.[487] Our second COVID Halloween saw a similar scene to the first, with eye masks rather than mouth and nose masks prevailing.

By Christmas 2021, COVID's upward curve had become a straight line.[488] Although Public Health proclaimed the need for another lockdown, our province's business-elected premier refused, even after Quebec's premier had locked down the province next door.[489] Our premier read from the business-friendly script that adhered to the philosophy that, as the Omicron variant is less likely to put persons in ICUs than previous variants, it will soon become just a nuisance virus like influenza.[490] Such philosophy forgets that seniors and other vulnerable persons can die from influenza, and ignores the mathematics that even if only a small percentage of Omicron patients are ill enough to be admitted to our ICUs, a small percentage of a large total will put many persons in ICU beds, not to mention coffins.[491] Our premier's business-based decision caused COVID to again overflow our ICUs.[492]

On February 16, 2022, our province's Public Health Officer reported 1,403 patients in hospitals with COVID, and that many of these patients would become "long-haulers"[493] suffering years of complications from having been infected by COVID. The term "long-

---

[487] Alhmidi, 2021; Lawless, 2021.
[488] Mitsui, 2021.
[489] Stevenson, 2021.
[490] Jabakhanji, 2022.
[491] COVID's very contagious Omicron variant would spread uncontrollably after that Christmas in 2021, and even though less lethal than previous variants, the vast number of infections caused many deaths among seniors and other vulnerable populations, as well as an epidemic of illness among the nurses and other hospital staff required to ensure essential healthcare ("Planning a holiday gathering? Here are COVID-19 rules and advice," 2021).
[492] "Ontario reports 1,829 patients hospitalized with COVID-19 and 435 in ICUs," 2022.
[493] The term "long-hauler" was introduced in relation to COVID 2020 (Siegelman, 2020), and echoed through 2021 (Callard & Perego, 2021)

hauler" became ironic when convoys of long-hauler antivaxxer truckers[494] occupied Canada's capital Ottawa for three weeks[495] and blocked border crossings to the United States at Windsor, Ontario,[496] Emerson, Manitoba,[497] and Coutts, Alberta.[498]

Loosening of restrictions in March 2022, again for the purpose of "keeping businesses economically sound,"[499] spiked another COVID wave, but our premier was not concerned, stating that our ICUs had adequate capacity to absorb more COVID patients. Our premier, by sentencing citizens to ICUs in the name of business, was sacrificing to terrible illness a portion of the population he was supposed to protect. At the beginning of May, parades of motorcyclists invaded Canada's capital; the rippling of their motors echoing the trucker convoys of February, and their messaging resonating with Canadians who favoured "personal freedom" over mandated vaccine.

A few years ago, I had the gift of visiting The Salk Institute in La Jolla, California, and absorbing the inspiration of the physician who contributed so much to humanity. Walking on the beach in front of The Salk Institute reminisced Sunnyside Beach in Toronto, where my Grandmother's fear that I would catch the polio virus social-distanced me. We owe it to Dr. Salk to help "the sun"[500] of vaccination

---

[494] I have no intention of dumping on truckers as my Grandfather was a trucker, and as a teenager I was blown away by John Steinbeck's *Grapes of Wrath* (Steinbeck, 1939), finding resonance with the part in which two long-hauler truckers surreptitiously pay for the candy canes admired by two young children fleeing the poverty of Oklahoma's dustbowl with their extended family.

[495] "The convoy crisis in Ottawa: A timeline of key events," 2022.

[496] Fraser, 2022.

[497] "Protesters continue to blockade major Canada-U.S. border crossing in Manitoba," 2022.

[498] "Frustration mounts as blockade snarling access to U.S. border continues at Alberta port of entry," 2022.

[499] Office of the Premier, 2021.

[500] As noted previously in this Chapter, Dr. Salk responded to the journalist who asked who owns the patent to the Salk vaccine with, "Could you patent the sun?" (Bos, 2013).

come to persons for whom it remains distant, whether the persons live in a privileged country like Canada, or in a socio-economically disadvantaged country. We owe it to Dr. Salk to help the sun of vaccination come, whether the vaccine is to prevent the COVID-19 virus, or the polio virus still prevalent in disadvantaged countries. We owe it to Dr. Salk to help the sun of vaccination come to assist Public Health Units in their efforts to limit COVID's spread, hospitalizations, and deaths. We owe it to Dr. Salk to help the sun come to end the COVID pandemic.

References

(2021, February 8). Homeowner slapped with hefty ticket as police sack Super Bowl bash. *London Free Press*. https://lfpress.com/news/local-news/homeowner-slapped-with-hefty-ticket-as-police-sack-super-bowl-bash

(2021, Dec 17). 'Omicron will not take a holiday': Ontario announces new limits for gatherings, businesses as COVID-19 spikes. Canadian Broadcasting Corporation News. https://www.cbc.ca/news/canada/toronto/covid-19-ontario-dec-17-2021-omicron-cases-1.6289571

(2021, Dec 4). Planning a holiday gathering? Here are COVID-19 rules and advice. Canadian Broadcasting Corporation News. https://www.cbc.ca/news/canada/ottawa/holiday-2021-covid-gathering-party-ottawa-gatineau-1.6272131

(2022, Jan 04). Coronavirus: What's happening in Canada and around the world on Jan. 4. Canadian Broadcasting Corporation News. https://www.cbc.ca/news/world/coronavirus-covid19-canada-world-jan4-2022-1.6303422

(2022, Jan 4). 2 *GTA hospitals declare 'code orange' as Ontario prepares to tighten public health measures.* Canadian Broadcasting Corporation News.

https://www.cbc.ca/news/canada/toronto/covid-19-ontario-jan-4-2022-hospitals-icus-1.6303596

(2022, Jan 29). *COVID-19 protesters demonstrate across Canada in support of truck convoy in Ottawa*. Canadian Broadcasting Corporation News. https://www.cbc.ca/news/canada/canada-protests-truck-convy-1.6332680

(2022, Jan 31). *Frustration mounts as blockade snarling access to U.S. border continues at Alberta port of entry*. Canadian Broadcasting Corporation News. https://www.cbc.ca/news/canada/calgary/blockade-coutts-alberta-trucker-covid-convoy-1.6333957

(2022, Feb 11). Ontario reports 1,829 patients hospitalized with COVID-19 and 435 in ICUs. CBC News. https://www.cbc.ca/news/canada/toronto/covid19-ontario-february-11-2022-1.6347806

(2022, Feb 17). *The convoy crisis in Ottawa: A timeline of key events*. Canadian Broadcasting Corporation News. https://www.cbc.ca/news/canada/ottawa/timeline-of-convoy-protest-in-ottawa-1.6351432

(2022, Feb 12). *Protesters continue to blockade major Canada-U.S. border crossing in Manitoba*. Canadian Broadcasting Corporation News. https://www.cbc.ca/news/canada/manitoba/protest-canada-us-border-blockade-manitoba-1.6349659

Alhmidi, M. (2021, Oct 22). *Ontario university vaccine mandates mean some students being barred from campus*. CTV News. https://toronto.ctvnews.ca/ontario-university-vaccine-mandates-mean-some-students-being-barred-from-campus-1.5634448

Andrews B & Anand A. (2022, Jan 31) After weekend of protests, Ottawa residents are feeling the effects. Canadian Broadcasting Corporation News. https://www.cbc.ca/news/canada/ottawa/convoy-workers-two-days-later-1.6333017

Ansari, S. (2020, Jul 7). *Pick our fruit, get COVID-19*. Maclean's. https://www.macleans.ca/opinion/coronavirus-exposing-canada-exploitative-immigration-practices/

Baicus, A. (2012). History of polio vaccination. *World Journal of Virology, 1*(4), 108–114.

Bicknell, B. (2020, Nov 10). *Host facing minimum $10k fine after large Halloween party near Western University campus*. CTV News. https://london.ctvnews.ca/host-facing-minimum-10k-fine-after-large-halloween-party-near-western-university-campus-1.5182432

Bicknell, B. (2021, Mar 31). *LHSC spared ICU onslaught…for now*. CTV News. https://london.ctvnews.ca/lhsc-spared-icu-onslaught-for-now-1.5370455

Bogdan, S. (2021, Dec 24). *Christmas 2021: What's open and closed this holiday season in London, Ont*. Global News. https://globalnews.ca/news/8472599/christmas-2021-open-and-closed-london-ont/

Bogart, N. (2020, Sept 23). *Killer parties: University students' get-togethers are putting lives at risk, officials warn*. CTV News. https://www.ctvnews.ca/health/coronavirus/killer-parties-university-students-get-togethers-are-putting-lives-at-risk-officials-warn-1.5110539

Bos, C. (2013). *Jonas Salk—"Could you patent the sun?"* Awesome Stories. https://www.awesomestories.com/asset/view/Jonas-Salk-Could-You-Patent-the-Sun-1

Butler, C. (2021, Apr 1). *Western University moves classes online, asks students in residence to go home*. Canadian Broadcasting Corporation News. https://www.cbc.ca/news/canada/london/western-university-moves-classes-online-asks-students-in-residence-to-go-home-1.5972923

Callard F, Perego E. How and why patients made Long Covid. Soc Sci Med. 2021 Jan;268:113426. doi: 10.1016/j.socscimed.2020.113426. Epub 2020 Oct 7. PMID: 33199035; PMCID: PMC7539940

Chatham, L. (2020, Sept 24). *UGA students face social stigma associated with contracting COVID-19.* The Red & Black. https://www.redandblack.com/uganews/uga-students-face-social-stigma-associated-with-contracting-covid-19/article_43e85402-fe5b-11ea-ae33-9753b4ab42c7.html

Daigle, T. (2022, Jan 15). *In this Ontario hospital, it's mostly the unvaccinated who are overwhelming the ICU.* Canadian Broadcasting Corporation News. https://www.cbc.ca/news/health/sarnia-bluewater-health-hospital-covid-patients-1.6315681

Denette, N. (2021, Mar 11). *Variants of concern cause more than 40% of new COVID-19 cases in Ontario, experts say.* Canadian Broadcasting Corporation News. https://www.cbc.ca/news/canada/toronto/covid-19-ontario-march-11-2021-variants-of-concern-1.5945381

Dubinski, K. (2020, Nov 2). *Western University "troubled" after large student Halloween party at this house.* Canadian Broadcasting Corporation News. https://www.cbc.ca/news/canada/london/western-university-troubled-after-large-student-halloween-party-at-this-house-1.5786310

Dubinski, K. & Rodriguez, S. (2021, Apr 13). *London, Ont., meat plant shut for 2 weeks amid COVID-19 outbreak affecting 82 workers.* (2021, April 13). Canadian Broadcasting Corporation News. https://www.cbc.ca/news/canada/london/cargill-meat-plant-covid19-temporary-shutdown-1.5985889

Favaro, A. & Jones, A. (2022, Jan 12). *Inside an ICU where 70 per cent of COVID-19 patients are unvaccinated.* CTV News. https://www.ctvnews.ca/health/coronavirus/inside-an-icu-where-70-per-cent-of-covid-19-patients-are-unvaccinated-1.5738198

Fraser, K. (2022, Feb 14). *Ambassador Bridge reopens with heavy police presence around former Windsor, Ont., protest site.* Canadian Broadcasting Corporation News.

https://www.cbc.ca/news/canada/windsor/ambassador-bridge-reopens-monday-1.6350729

Government of Canada. (updated 2022 Feb 15). Physical distancing: How to slow the spread of COVID-19. https://www.canada.ca/en/public-health/services/publications/diseases-conditions/social-distancing.html

Government of Canada. (2018). *Poliomyelitis (Polio): For health professionals.* https://www.canada.ca/en/public-health/services/diseases/poliomyelitis-polio/health-professionals.html

Graham, A. (2021, Mar 17). *Coronavirus: Patios busy, student neighbourhoods quiet on St. Patrick's Day in London, Ont.* Global News. https://globalnews.ca/news/7703616/st-patricks-day-london-ont-2021-coronavirus/

Graham, A. (2021a, Apr 23). *COVID-19: Cargill reopens following 2nd largest workplace outbreak in London, Ont.* Global News. https://globalnews.ca/news/7783304/covid-19-london-cargill-reopens/

Guse, J. (2022, Jan 10). *Potentially growing COVID-19 outbreak at High River Cargill meatpacking plant.* Global News. https://globalnews.ca/news/8499422/covid-19-outbreak-high-river-cargill-meat-packing-plant/

Hristova, B. (2020, Sept 30). *This student has COVID-19 — and the stigma that comes with it.* Canadian Broadcasting Corporation News. https://www.cbc.ca/news/canada/hamilton/student-has-covid-19-1.5743179

Hotez, P. (2021). COVID vaccines: time to confront anti-vax aggression. *Nature,* 592(7856), 661. https://doi.org/10.1038/d41586-021-01084-x

Jabakhanji, S. (2022, Jan 3). Ontario moves school online, closes indoor dining and gyms as part of sweeping new COVID-19 measures. CBC News. https://www.cbc.ca/news/canada/toronto/covid-19-ontario-jan-3-2022-ford-public-health-measures-1.6302531

Khandia, R., Singhal, S., Alqahtani, T., Kamal, M.A., El-Shall, N.A., Nainu, F., Desingu, P.A., Dhama K. (2022). Emergence of SARS-CoV-2 Omicron (B.1.1.529) variant, salient features, high global health concerns and strategies to counter it amid ongoing COVID-19 pandemic. *Environ Res.* doi: 10.1016/j.envres.2022.112816. Epub ahead of print. PMID: 35093310; PMCID: PMC8798788.

Kurjata, A. (2021, Oct 22). *Husband regrets anti-vaxx stance as wife lies in a coma 800 km from home.* Canadian Broadcasting Corporation News. https://www.cbc.ca/news/canada/british-columbia/anti-vaccine-fort-st-john-pregnant-wife-1.6222325

Lawless, J. (2021, Sept 1). *Group of Queen's University students start petition against vaccine mandate.* Global News. https://globalnews.ca/news/8159609/queens-university-students-petition-vaccine-mandate/

Lee-Shanok, P. (2021, Sept 3). *Protesters against COVID-19 vaccine mandates say they're pro-freedom, 'not anti-vax'.* Canadian Broadcasting Corporation News. https://www.cbc.ca/news/health/protesters-covid-vaccine-mandates-infringe-freedoms-1.6164308

Marchand, L. (2022, Jan 5). *Passengers on Sunwing party plane could face jail time, thousands in fines.* Canadian Broadcasting Corporation News. https://www.cbc.ca/news/canada/montreal/sunwing-cancun-flight-1.6304854

Mendoza, J. (2022, Feb 4). *COVID-19 vaccine immunization development in Mexico 2022.* https://www.statista.com/statistics/1104529/population-vaccinated-against-covid-19-mexico/

Mitsui, E. (2021, Dec 6). *Ontario reports 887 new COVID-19 cases as ICU admissions climb to highest level in 2 months.* Canadian Broadcasting Corporation News. https://www.cbc.ca/news/canada/toronto/covid-19-ontario-dec-6-2021-cases-icu-admissions-up-1.6274971

Neustaeter, B. (2020, Apr 29). *These are the meat plants in Canada affected by the coronavirus outbreak.* CTV News. https://www.ctvnews.ca/health/coronavirus/these-are-the-meat-plants-in-canada-affected-by-the-coronavirus-outbreak-1.4916957

Office of the Premier. (2021, Apr 7). *Ontario enacts provincial emergency and stay-at-home order* [News release]. https://news.ontario.ca/en/release/61029/ontario-enacts-provincial-emergency-and-stay-at-home-order

Paling, E. (2021, Dec 24). *Omicron's prevalence should shake off COVID-19's lingering stigma, experts say.* Canadian Broadcasting Corporation News. https://www.cbc.ca/news/health/omicron-should-shake-off-covid-stigma-1.6296669

Pelley, L. (2022, Jan 26). Omicron subvariant BA.2 raises new questions about puzzling evolution of virus behind COVID-19. Canadian Broadcasting Corporation News. https://www.cbc.ca/news/health/omicron-subvariant-ba-2-raises-new-questions-about-puzzling-evolution-of-virus-behind-covid-19-1.6327270.

Phillips, K. (2021, Feb 12). *Roberta Place long-term care home in Barrie, Ont. reports 70th resident death amid COVID-19 outbreak.* CTV News. https://barrie.ctvnews.ca/roberta-place-long-term-care-home-in-barrie-ont-reports-70th-resident-death-amid-covid-19-outbreak-1.5307406

Rashid, H, Khandaker, G, & Booy, R. (2012). Vaccination and herd immunity: What more do we know? *Current Opinion in Infectious Diseases, 25*(3), 243–249.

Rivers, H. (2020, June 9). *COVID-19: London doctor speaks out about stigma of having illness.* London Free Press. https://lfpress.com/news/local-news/covid-19-london-doctor-speaks-out-about-stigma-of-having-illness

Rocca, R. (2020). Ontario to enter 'provincewide shutdown' on Boxing Day. Global News.

https://globalnews.ca/news/7535230/ontario-new-coronavirus-shutdown-boxing-day/

Rutty, C.J., Barreto, L., Van Exan, R., Gilchrist, S. (2005). *Conquering the crippler: Canada and the eradication of polio.* Canadian Public Health Association. https://immunize.ca/sites/default/files/resources/107e.pdf

Sharkey, J. (2021, April 16). *What you need to know about London's ICU capacity as COVID cases surge.* Canadian Broadcasting Corporation News. https://www.cbc.ca/news/canada/london/london-health-sciences-centre-lhsc-pandemic-1.5990478

Siegelman JN. Reflections of a COVID-19 Long Hauler. JAMA. 2020 Nov 24;324(20):2031-2032. doi: 10.1001/jama.2020.22130. PMID: 33175108

Serkis, A. (Director). (2017). *Breathe.* Bleecker Street.

Steinbeck, J. (2014). Grapes of wrath. Viking. (Original work published 1939)

Stevenson, V. (2021, Dec 30). *Quebec imposes curfew, closes restaurant dining rooms and further delays return to school.* CBC News. https://www.cbc.ca/news/canada/montreal/quebec-news-conference-new-year-public-health-measures-1.6300455

Stone, L. & Moore, O. (2021, Mar 26). *Ontario loosens COVID-19 restrictions even as concern over rising case counts, hospital admissions grow.* The Globe and Mail. https://www.theglobeandmail.com/canada/article-ontario-loosens-covid-19-restrictions-even-as-concern-over-rising-case/

Webster, P. (2021). COVID-19 highlights Canada's care home crisis. *Lancet, 397*(10270), 183.

Chapter 20

Webinar Physicians' Cavalier Terms for COVID-Ventilator Triage of Disabled Persons

Webinar physicians' cavalier triage terms
Spread more than contagion of COVID-ventilator restriction
Their terms set a dangerous precedent
For ventilator-restriction from persons with disabilities;
Persons seen as "less healthy"
Thus more worthy of death with dignity[501]
"If two MDs agree"[502] that's all you need
To be a hanging judge[503] and jury in Canada.

---

[501] See Chapter 17, Ruth.

[502] Quotation marks indicate exact terms used by physicians in the Webinar on COVID-ventilator restriction. Unfortunately, the Webinar did not provide the ability to cite or reference exists not only in healthcare journals, but in the lay press. This inability makes webinars potentially dangerous in healthcare.

[503] A "hanging judge" is the deprecating term for a judge who prefers "hang by the neck until dead" rather than consider a life-sentence in prison.

Webinar physicians' cavalier triage terms
Limit persons to just numbers
Entered in "clinical factor calculators,"
And coloured on triage-criteria charts;
Charts that set persons of difference apart
For better "return on investment,"
And spread the virus of the myth
Of "finite health resources."

Webinar physicians' cavalier triage terms,
Such as "we can't afford to have this,"
Must be rebutted with our insist
"That we can't afford not to have this";
As persons are not cash to be transacted,
Nor liquid assets to be liquidated,
Nor currency to be contemporarily based
On "predicted short-term mortality."

Webinar physicians' cavalier triage terms
Dissolve physician trust in hemlock,[504]
And our profession in suspicion,
Confirming the worst opinion of physicians;
Thus "if two MDs agree" with triage
We must ensure they are not one God,
As physicians must not be omnipotent
Purveyors of COVID-ventilator restriction.

---

[504] Physicians were not always trusted because of their equal knowledge of potions and poisons, and the participation of some physicians in assassinations.

Webinar physicians' cavalier triage terms
Spread concern that the physicians who use them
Are insensitive to persons with disabilities,
And other persons of difference;
For though ventilator-restriction is "still hypothetical"
Hypothetical too often masks prophetical
Of the death knell of those deemed less fit,[505]
Of the death knell of justice in our health system.

A preliminary version of this poem was published in *Impact Ethics* on May 13, 2021, yet the threat of COVID-ventilator restriction from persons with disabilities continues into 2022.

References

Dawidowicz L, (1975). The war against the Jews 1933-1945. Holt, Rinheart & Winston; New York, 1975

---

[505] Hitler used the language of "less fit" and "less human" to make it easier for Germans and persons in other countries to accept the extermination of Jews (Dawidowicz, 1975).

Chapter 21

COVID Injustice Before I Heard the Word "COVID"

Not until March 2020 did I learn of the "virus of concern." Even though I am a physician. Even though I had developed pneumonia from the virus two weeks before. Even though the virus was rampant in Wuhan, China, since autumn.[506] The Chinese government had been successful in suppressing knowledge of the virus, but not the virus itself.

In order to accomplish this suppression of knowledge, the Chinese government "banned social-media posts on the virus, stopped symptomatic people from entering hospitals, unleashed a stream of TV propaganda downplaying its severity,"[507] and even punished doctors who spoke of the virus.[508] By the time the Chinese government acknowledged the virus, Pandora's box[509] had been open long enough for the virus to plethora the world.

---

[506] World Health Organization, 2021.
[507] Zeitchik, 2021.
[508] Zeitchik, 2021.
[509] According to Greek mythology, Pandora's box contained all the world's evil, and thus must never be opened. Eventually Pandora could not resist opening the box, and waves of evil spread through the world.

Suppression of knowledge of the virus permitted unmasked air travel to continue unabated around the globe, including the Sunwing flight to Mexico for Reading Week[510] that incubated my partner and me, and likely the virus. On our third morning in the sun, I succumbed to headache, fever, coughing, and difficulty breathing. That afternoon I conceded to my symptoms, not to mention my partner's insistence, and asked directions to a local physician. He tapped my chest, listened to my lungs, gently told me I had pneumonia, and prescribed antibiotics. The following day my shortness of breath had so dramatically worsened that my partner found a seat for me on the next plane back to Canada.

On my flight to Canadian healthcare, there was little room to "social distance" before I learned the term "social distance." The best I could do was press my fevered forehead against the cold window on my right, trying to keep my increasingly short breath away from the passenger on my left, fortunately an empty seat away. I asked the flight attendant for a mask, but as it was still rare to wear a mask, she could not find one for me. I coughed dutifully into my wool sweater's elbow, while at the same time trying to infuse air into my fluid-filled lungs. I felt like I was drowning.[511]

Eventually I needed to pee, and I struggled from my seat, trying not to touch the top of the seats of the row in front, while murmuring apologies behind my sleeve to the woman sitting between me and the aisle. Out of breath and dizzy, I wobbled down the long aisle to the bathroom, cognisant of staying as far from other passengers as possible. Turbulence forced me to touch the top of one of their seats, which I quickly Purelled with a squirt from the small bottle I keep in my pocket proximate to my passport. I used up the rest of the bottle on the bathroom's door handle and sliding lock, and was happy to see

---

[510] "Reading Week" is the ironic misnomer for the fun-break in February that universities afford students (and professors), presumably to catch up on course material.

[511] I describe my fear of drowning in Chapter 3, I'm Sorry Ronnie. To obviate this fear I swim long distances in open water, which was the reason I was in Mexico for Reading Week.

an almost-full pink soap pump on the sink. After returning the seat to its down position, I used the pink soap effusively on the toilet's flush handle and seat. I then soaped the soap pump, the sliding lock, and the interior door handle, and exited.

After the jet landed, I waited for the other passengers to disembark into our small terminal, and followed at a more-than-six-feet distance, before the insistence on a six-feet distance was proclaimed.[512] In the accordioned passageway, I was careful not to pant on the security guard as I huffed by her in the narrow corridor to Customs, where I plastered myself on a warm wall while waiting for the other passengers to funnel the slits between booths. Another security guard, on observing my hesitancy to go through Customs, asked if there was a problem. I responded, "I have pneumonia, and I don't want to infect other passengers, or you for that matter." After Customs, I crouched in a corner of Baggage Retrieval until I spotted my bag patiently descending its designated conveyor belt. I took a deep breath, pounced on my bag, exited Baggage Retrieval, and exhaled. I kept as far from others as possible as I zig-zagged toward the terminal's exit door, and dashed into winter. I tried to keep my distance as I felt deflated by the long lineup at the taxi stand that seemed to never melt, partially for the reason that my distance permitted other passengers to keep cutting in.

When I finally stepped into a cab, I asked the driver to lower all the windows. He resisted amidst the winter cold, but after hearing my first cough quickly complied. My shortness of breath seemed to increase as the taxi neared my home, and I debated requesting the driver continue three more blocks to University Hospital; however, my fear my colleagues there would insist on a ventilator caused me to reconsider. My claustrophobia fears a ventilator more than I fear death,[513] as I had made peace with death three years previous because of metastatic cancer.[514]

---

[512] Government of Canada, updated Feb 15, 2022.
[513] See Claustrophobia's Fear of a COVID Ventilator (Nisker, In press 2022).
[514] See Chapter 25, The Arrogance of "But All You Need Is a Good Index Finger."

I handed the driver a twenty-dollar bill at arm's length over his shoulder, told him to keep the change, and suggested he Purell the bill along with any other surfaces I might have touched. I grabbed my bag and briefcase from the seat beside me, stumbled out of the cab, and wobbled up the walk. Dizziness made it difficult for my key to find the front door's lock, and after it did and turned, I fell through the door to face-plant on the front hall floor. I tried to make it up the five split-level steps to the bedroom, but on the third fell to my knees gasping. I took out my cell and pressed 911, but not the call button. Even though I felt like I was drowning,[515] my fear of a ventilator remained greater than my fear of death.[516] With my thumb still cautiously on the call button, I crawled up the last two steps and collapsed on the upstairs carpet. Dizziness inhibited further progress to bed, and the headache would keep me awake anyway.

The next morning my partner called to check on me. Although I insisted she stay in Mexico for the rest of Reading Week, she caught the next flight home. She had just made it back when she developed headache, dizziness, cough, and fever. Fortunately, her non-chemoed immune system is stronger than mine, and she did not develop full-fledged pneumonia. For the next two days, I still felt like I was drowning, but whatever bug this was, it was not putting me in the Hospital. The following week I would learn that this bug was a "virus of concern."

The Report of the World Health Organization's global study of the origins of SARS-CoV-2 (COVID-19), which assessed "surveillance data on all-cause mortality and pneumonia-specific mortality from Wuhan city and the rest of Hubei Province,"[517] found that samples from patients with exposure in the Huanan market in December 2019 confirmed a cluster there.[518] The WHO report went on to suggest that

---

[515] I reflect on my young cousin's drowning in Chapter 3, I'm Sorry Ronnie.
[516] See Claustrophobia's Fear of a COVID Ventilator (Nisker, In press 2022).
[517] World Health Organization, 2021.
[518] World Health Organization, 2021.

the outbreak started in autumn 2019,[519] and cautioned that many other cases had been associated with other markets in Wuhan in which live animals known to carry SARS-CoV-2 are sold, including pangolins, mink and cats.[520]

I contend that the Chinese government's six-month camouflage of knowledge of the virus that would be called COVID-19 is the greatest injustice in medicine perpetrated in my lifetime. By May 2022, there were over 15 million COVID deaths worldwide,[521] and at least 40,843 COVID-19 deaths in Canada.[522] This injustice of suppression of knowledge of a dangerous virus must never happen again, but will happen again unless opaque countries like China become transparent regarding their plagues before those plagues cross borders. This injustice will happen again unless public health agencies in all countries are empowered with the resources for vaccination to prevent the morphing of COVID sub-variants, like the Omicron sub-variants that are part of our present,[523] and the new COVID sub-variants that will become part of our future.[524] More resources must be distributed to vaccinate vulnerable populations prone to COVID deaths, not only in "developed" countries like Canada,[525] but in socio-economically disadvantaged countries. More resources must be distributed to ensure that global access to new vaccines extinguishes the emergence of new COVID variants, and other viruses a soon as they become apparent.

---

[519] World Health Organization, 2021.
[520] World Health Organization, 2021. Bats are the likely vector of COVID from forests to domestic animals.
[521] "Pandemic death toll at end of 2021 may have hit 15 million people, WHO estimates", 2022.
[522] John Hopkins University of Medicine, 2022.
[523] The evolving Omicron subvariants (Khandia, 2022) are more resistant to our vaccines (Dolgin, 2022).
[524] A new Omicron subvariant is spreading rapidly at the end of March 2022 (Patterson, 2022).
[525] Brown et al, 2022.

References

(2022, Apr 5). COVID outbreak 'extremely grim' as Shanghai extends lockdown. CP24 News. https://www.cp24.com/world/covid-outbreak-extremely-grim-as-shanghai-extends-lockdown-1.5848503

(2022, May 5). Pandemic death toll at end of 2021 may have hit 15 million people, WHO estimates. Canadian Broadcasting Corporation News. https://www.cbc.ca/news/health/who-excess-death-modelling-1.6442146

Brown, H. K., Saha, S., Chan, T., Cheung, A. M., Fralick, M., Ghassemi, M., Herridge, M., Kwan, J., Rawal, S., Rosella, L., Tang, T., Weinerman, A., Lunsky, Y., Razak, F., & Verma, A. A. (2022). Outcomes in patients with and without disability admitted to hospital with COVID-19: a retrospective cohort study. *Canadian Medical Association Journal*, 194(4), E112–E121. https://doi.org/10.1503/cmaj.211277

Cabrera, H. (2022, Apr 18). Sunwing flights delayed, passengers stranded amid network-wide issue. Canadian Broadcasting Corporation News. https://www.cbc.ca/news/canada/montreal/sunwing-network-wide-issue-flights-delayed-1.6422610

Dolgin E. (2022). Omicron thwarts some of the world's most-used COVID vaccines. Nature, 601(7893), 311. https://doi.org/10.1038/d41586-022-00079-6

Government of Canada. (updated 2022 Feb 15). Physical distancing: How to slow the spread of COVID-19. https://www.canada.ca/en/public-health/services/publications/diseases-conditions/social-distancing.html

John Hopkins University of Medicine. (2022, Feb 16). COVID-19 Dashboard. John Hopkins University of Medicine. https://coronavirus.jhu.edu/map.html

Khandia R, Singhal S, Alqahtani T, Kamal MA, El-Shall NA, Nainu F, Desingu PA, Dhama K. (2022). Emergence of SARS-CoV-2 Omicron (B.1.1.529) variant, salient features, high global health concerns and strategies to counter it amid ongoing COVID-19 pandemic. Environ Res. doi: 10.1016/j.envres.2022.112816. Epub ahead of print. PMID: 35093310; PMCID: PMC8798788.

Nisker, J. (In press 2022). Confined to the COVID Sidelines: New and Selected Verses.

Patterson D. (2022, Mar 23). Doctors say Omicron subvariant a sign 'pandemic is not over yet'. Canadian Broadcasting Corporation News. https://www.cbc.ca/news/canada/saskatchewan/omicron-subvariant-ba2-1.6393804

Reuters T. (2022 Mar 27). China announces citywide lockdown of Shanghai in 2 stages as COVID-19 cases spike. Canadian Broadcasting Corporation News. https://www.cbc.ca/news/world/shanghai-covid-19-lockdown-two-stages-march-27-1.6399232

World Health Organization. WHO-convened global study of origins of SARS-CoV-2: China Part [Internet]. World Health Organization. [2021, 14 January-10 February]. Available from: https://www.who.int/publications/i/item/who-convened-global-study-of-origins-of-sars-cov-2-china-part

Zeitchik S. (2021, Jan 28). A scathing new documentary from HBO alleges a Chinese coverup on the coronavirus. Washington Post. https://www.washingtonpost.com/business/2021/01/28/china-hbo-covid-film/

Chapter 22

COVID Aggression Condemns a Muslim Family Near Our Medical School[526]

Post-chemo's fear of a COVID-ventilator
Cloistered me from contagion in social-distance shelters
Until tragedy found our university town
And unbound my COVID shackles to shoulder-to-shoulder kneel
    down
On blood-stained ground where a family had just been murdered
At the hate-sharpened end of a tire-skid scythe of black rubber
Where I throw blossoms from my garden on a flower volcano's eruption
And join hands with new friends who just happen to be Muslim

---

[526] Western University is in London Canada, a few blocks from where this family was intentionally struck by a truck on June 6, 2021, and where the mother and father in this family achieved Master's degrees in physiotherapy. In the three months following the attack on this family, COVID-aggression was again exhibited in four sexual assaults (Dubinski, 2021), and a manslaughter on Western's campus (Dolynny, 2021). I have also observed COVID-aggression when the cars and trucks of young men speed around the corner separating Western University from my home.

Their families may have come from Lebanon or Pakistan
Or Saudi Arabia or Africa or Afghanistan or Kuwait
For a fate for women and children free of misogynistic hate
Did they now contemplate their migrate to Canada a mistake
And closer friends had come as "foreign medical graduates"[527]
To learn more medicine in order to become specialists
Did they now sense concern of appearing a bit foreign
Like they did after 9/11 because they just happen to be Muslim?

In our university town's multi-ethnic and educated fabric
It was easy to forget that beneath the surface white supremacists lurk
Even as we watched their horrific prejudice in Alberta and Quebec
And other provinces where some citizens for Muslims lack respect
Indeed are circumspect of all persons whose appearance is different
Whether the persons are Indigenous or Black or immigrant
Or living with disabilities or gay or lesbian or trans
Along with the persons who just happen to be Muslim

The next day I entered the hospital where for many years I have practiced
And where Intensive Care has expanded with the COVID-virus infected
As well as a nine-year-old boy with fractures and organ damage
From being plowed down by the hate-virus that infected a white supremacist
Whose belief that killing Muslims would be so well received
He didn't need a balaclava nor white-hooded robe to attack a family
On their evening walk near our University where the boy's parents had studied
In a country they thought was free even if they just happen to be Muslim.

---

[527] Pejoratively called "FMGs."

How do you tell a nine-year-old in an intensive care bed
When he asks where are his parents that his parents are dead
How do you submerge the murders of his sister and grandmother
Until he recovers enough to absorb what the atrocity that occurred
How do you disguise from the eyes of a child full of goodness
That his family was murdered by hate purchased on the net
How can this child look forward to a future without the concern
That murder will capture him for the sin he just happens to be
    Muslim

References

Dolynny, T. (2021). Second person wanted in connection with Western student's death. CBC News. https://www.cbc.ca/news/canada/london/second-person-wanted-in-connection-with-western-student-s-death-1.6180684

Dubinski K. (2021, Sept 13). Arrests made after 4 Western students reported sexual assaults in past week, university official says. Canadian Broadcasting Corporation News. https://www.cbc.ca/news/canada/london/western-campus-sexual-violence-reports-1.6173443

Chapter 23

The Lottery

In the midst of COVID January 2021, the Canadian Broadcasting Corporation reported Toronto's intensive care physicians were considering a triage or even "lottery" system for admission to this care.[528] The physicians defended their position with the point that critical care medicines were "in short supply." Their suggestion of a "lottery" for who will live and who will die in 2021 was horrifyingly reminiscent of Shirley Jackson's short story "The Lottery," written in 1948.[529]

Healthcare lotteries in Canada are many and massive, and they help fund hospital constructions and operations that eventually promote better patient care. However, the less altruistic and more magnetic attraction to pick up lottery tickets is the "jackpot": a Muskoka cottage or the equivalent in millions of dollars. I contend that intensive care physicians' proposal of a "lottery" for the jackpot of living is different and disturbing, and preminisces the prejudice against socio-economically disadvantaged persons that will occur with a lottery system for hospital admission.[530]

---

[528] Jabakhanji, 2021.
[529] Jackson, 1948.
[530] Mykitiuk & Lemmens, 2020.

In Jackson's "The Lottery," residents of a quiet late-1940s American town draw paper slips from a metal box, the purpose of which the reader has no knowledge of until the last paragraph. The anticipation of the "lottery" in Shirley Jackson's story is celebratory, and the day of the "lottery" seems like a community picnic. Halfway through the story, Mrs. Delacroix says to Mrs. Graves, "Seems like there's no time at all between lotteries any more," a prescient observation relevant to the threat of a ventilator-lottery system in 2022. Mrs. Delacroix's comment conveys my concern that if we use a lottery in 2022 for access to intensive care medications and ventilators, what other critical aspects of our society will become just as cavalierly[531] restricted?

The celebratory attitude of the citizens in "The Lottery" continues to the story's ending; that is, except for the concern of the person whose paper name was just drawn from steel box, "Tessie Hutchinson."

The children had stones already, and someone gave little Davy Hutchinson a few pebbles.

Tessie Hutchinson was in the center of a cleared space by now, and she held her hands out desperately as the villagers moved in on her. "It isn't fair," she said.

A stone hit her on the side of the head.

In addition to reminiscing Shirley Jackson's 1948 "The Lottery" the concept of a "lottery" for "crucial intensive care medications"[532] proposed in Canada January 2021 also reminisces the article by Diego Silva in the prestigious medical journal *Chest*[533] at the

---

[531] See Chapter 20, Webinar Physicians' Cavalier Terms for COVID-Ventilator Triage of Disabled Persons.
[532] Canadian Broadcasting Corporation News, January 19, 2021.
[533] Silva, 2020.

beginning of the pandemic, in which he contends that "ventilators by lottery" is "the least unjust form of allocation in the corona disease 2019 pandemic."[534] Silva suggests that this is the majority view, and that of the scholars he references,[535] across much of bioethics.[536] However, I have the minority view, and contend that Silva's view is problematic, especially in wealthy countries like Canada, as it sacrifices individual persons to a lack-of-ventilator death; a death that approximates drowning.[537] I contend that rather than permitting drowning deaths in the name of inadequate health resources, we must demand an increase in resources to provide equal access to the best healthcare for all persons in our country, even if it means an increase in our taxes. Silva, I would suggest, agrees with me when he draws attention to the fact that utility arguments like his privilege the socio-economically well-off and contribute to further social inequalities.[538]

In one of the world's wealthiest countries, it is unethical and unnecessary to contemplate a "Lottery" for ventilators, as a small increase in our taxes can purchase as many ventilators as we need, as well as the training and funding of staff to operate them, obviating any need of a lottery system for access to critical care. I contend any consideration of a "Lottery" in healthcare is in itself unethical, and will lead to dehumanization of vulnerable populations such as persons with disabilities, and persons with socio-economically disadvantage. A person's life will always be worth more than a cottage or a million-dollar jackpot.

---

[534] Silva, 2020.
[535] Mounk, 2020; Dr. Ross Upshur was one of my mentors during my PhD studies at the University of Toronto; Silva, 2020.
[536] Silva, 2020.
[537] See Chapter 3, I'm Sorry Ronnie.
[538] Silva, 2020.

References

Jackson S. (1948). The Lottery. The New Yorker.

Mounk Y. (2020, Mar 11). *The Extraordinary Decisions Facing Italian Doctors.* The Atlantic. https://www.theatlantic.com/ideas/archive/2020/03/who-gets-hospital-bed/607807/

Emanuel EJ, Persad G, Upshur R, et al. Fair allocation of scarce medical resources in the time of COVID-19. *N Engl J Med.* 2020;382(21):2049-2055.

Jabakhanji S. (2021, Jan 14). ICU doctor calls triage protocol 'morally distressing' as province sees continued stress on care units. Canadian Broadcasting Corporation News. https://www.cbc.ca/news/canada/toronto/icu-care-triage-ontario-doctors-icu-1.5873664

Mykitiuk R, Lemmens T. (2020, Apr 19). Assessing the value of a life: COVID-19 triage orders mustn't work against those with disabilities. Canadian Broadcasting Corporation News. https://www.cbc.ca/news/opinion/opinion-disabled-covid-19-triage-orders-1.5532137

Silva D. S. (2020). Ventilators by Lottery: The Least Unjust Form of Allocation in the Coronavirus Disease 2019 Pandemic. *Chest, 158*(3), 890–891. https://doi.org/10.1016/j.chest.2020.04.049

Silva DS, Nie J, Rossiter K, Sahni S, Upshur REG. Contextualizing ethics: ventilators, H1N1 and marginalized populations. *Healthc Q.* 2010;13(1):32-36.

Silva DS, Smith MJ, Upshur REG. Disadvantaging the disadvantaged: when public health policies and practices negatively affect marginalized populations. *Canadian J Public Health.* 2013;104(5):410-412.

Chapter 24

Antivaxxer Xenophobic COVID Violence

Antivaxxer violence from defiance of vaccine
Condemns innocent persons to ventilator prisons
Even death sentences from COVID and its variants
Because these innocents believe the anti-vaccine propaganda they see

Antivaxxers spread contagion[539] of opposition to public health regulation
Indeed all mandates put in place to keep vulnerable persons safe[540]
And curb the surge of COVID's next scourge
Of radiation through of our post-restriction nation[541]

---

[539] The prescient film *Contagion* (2011), directed by Steven Soderbergh, has Kate Winslet portraying a public health physician who uncovers a lethal virus spreading in the Chicago region. This physician dies of the virus after taking swabs from numerous patients.

[540] An increase in COVID deaths and morbidities was reported for persons with disabilities in Canada (Brown et al, 2022).

[541] In April 2022, subsequent to the loosening of restrictions, a "sixth" wave of COVID began killing Canadians (Canadian Broadcasting Corporation News).

Antivaxxers are clogging our intensive care units[542]
"Occupying"[543] 70 percent of "expensive care" beds
Bed-blockers[544] of persons with no choice but to be there
With or without COVID who deserve the best care

Antivaxxers' venomous rhetoric super-spreads[545] misinformation
On the side-effects of vaccination and its mandate motivation
Selfishness sublimating any sniff of community consciousness
Violent verbiage "trumping"[546] any logic of vaccination

---

[542] More than 70% of persons in Canada's ICUs were unvaccinated in January 2022 (Daigle, 2022; Favaro & Jones, 2022).

[543] "Occupy" was proclaimed by the antivaxxers taking over the streets in Ottawa, Canada's capital, from Jan 28 to Feb 20, 2022, in a manner similar to when President Trump encouraged Americans to "occupy" the capital building of the Congress of the United States (Reston & Liptak, 2021). The Occupy movement in Toronto and other Canadian cities protested the global financial system, as well as economic and social inequality, and financial greed (Habib, 2011).

[544] "Bed blocker" (Styrborn & Thorslund, 1993; McGrail et al, 2001; Green, Croskerry & Rieck, 2020) is the pejorative term too-often used by physicians for persons whose likelihood of coming out of the ICU with "quality of life" is limited. See Chapter 20, Webinar Physicians' Cavalier Terms for Triage from COVID Ventilators; Also see Chapter 17, Ruth.

[545] "Super spreader" is term used of gatherings beyond the maximum limits that risk COVID-contagion from person to person (Kolnes et al, 2022; Li et al, 2022).

[546] I use the term "trumping" to convey former US president Donald Trump's encouragement of the right-wing assault to "occupy" the Congress of the United States on January 6, 2021, in which at least seven persons perished, and because on February 24, 2022, former president Trump condemned Canada's COVID restrictions (Canadian Broadcasting Corporation News). In addition, Trump supporters cheered the Canadian-trucker convoys opposing our vaccine mandate ("Truck convoy against COVID-19 public health measures rolls through Aylmer, Ont" 2022; Tasker, 2022; "COVID-19 protesters demonstrate across Canada in support of truck convoy in Ottawa", 2022); "Trumping" is a term used in playing card games

Antivaxxers proudly see themselves as courageous risk-takers
But the lives they courageously risk are often the lives of others,
Including nurses and doctors pushed past our previous limits
Because hours of work are not infinite nor steps to protect us from
    COVID

Antivaxxer-truckers are clogging our streets[547]
Blatantly blaring air horns of right-wing beliefs
Seen in their brazen waving of swastika rags
And Confederate flags of slavery past and KKK hate to this day[548]

Antivaxxer-truckers air-horn their dysphonic refrain
In convoys of false confidence across our nation
Enlisting their minions of similar "freedom"-minded Canadians[549]
Who bear the similar arrogance that condemns vulnerable citizens

---

like bridge when cards of one of the four suits (e.g., diamonds) become endowed with the power to be superior to other suits.

[547] "Freedom"; "Truck convoy against COVID-19 public health measures rolls through Aylmer, Ont," 2022; Tasker, 2022; "COVID-19 protesters demonstrate across Canada in support of truck convoy in Ottawa," 2022.

[548] The "Stars and Bars" was the Confederate flag in the American Civil War in which southern pro-slavery advocates of continuation proclaimed "freedom" of the "Confederacy" from President Lincoln's anti-slavery Emancipation Declaration on January 1, 1863. This flag proudly flew over the State Legislature Building in South Carolina until it was finally lowered in 2021. Graham, Mar 17, 2021.

[549] "Freedom"; "Truck convoy against COVID-19 public health measures rolls through Aylmer, Ont," 2022; Tasker, 2022; "COVID-19 protesters demonstrate across Canada in support of truck convoy in Ottawa," 2022.

Antivaxxer-truckers[550] take hostages on highways and border
    crossings[551]
In the name of the glorious purpose of personal "freedom"[552]
But what antivaxxers mean when they loudly espouse "freedom"
Is "freedom" of persons with similar belief systems

Antivaxxer arrogance leads many Canadians to believe
That opposition to vaccination is an acceptable alternative
To accepting community consciousness and moral responsibility
To protect others from disease no matter what doctrine we believe

References

(2022, Jan 29). *COVID-19 protesters demonstrate across Canada in support of truck convoy in Ottawa.* Canadian Broadcasting Corporation News. https://www.cbc.ca/news/canada/canada-protests-truck-convy-1.6332680

(2022, Feb 3). *Truck convoy against COVID-19 public health measures rolls through Aylmer, Ont.* Canadian Broadcasting Corporation News. https://www.cbc.ca/news/canada/london/truck-convoy-against-covid-19-public-health-measures-rolls-through-aylmer-ont-1.6338467

Andrews B & Anand A. (2022, Jan 31) *After weekend of protests, Ottawa residents are feeling the effects.* Canadian Broadcasting

---

[550] In no way is this poem intended to sully the reputation of truckers, for who I have had admiration and respect since I read John Steinbeck's "The Grapes of Wrath" (1939) as a young teen, not to mention the fact that my Grandfather with whom we lived when I was younger was a trucker.
[551] Lord, 2022.
[552] "Freedom"; "Truck convoy against COVID-19 public health measures rolls through Aylmer, Ont," 2022; Tasker, 2022.

Corporation News. https://www.cbc.ca/news/canada/ottawa/convoy-workers-two-days-later-1.6333017

Brown, H. K., Saha, S., Chan, T., Cheung, A. M., Fralick, M., Ghassemi, M., Herridge, M., Kwan, J., Rawal, S., Rosella, L., Tang, T., Weinerman, A., Lunsky, Y., Razak, F., & Verma, A. A. (2022). Outcomes in patients with and without disability admitted to hospital with COVID-19: a retrospective cohort study. *Canadian Medical Association Journal*, 194(4), E112–E121. https://doi.org/10.1503/cmaj.211277

Daigle T. (2022, Jan 15). *In this Ontario hospital, it's mostly the unvaccinated who are overwhelming the ICU.* Canadian Broadcasting Corporation News. https://www.cbc.ca/news/health/sarnia-bluewater-health-hospital-covid-patients-1.6315681

Favaro A & Jones A. (2022, Jan 12). *Inside an ICU where 70 per cent of COVID-19 patients are unvaccinated.* CTV News. https://www.ctvnews.ca/health/coronavirus/inside-an-icu-where-70-per-cent-of-covid-19-patients-are-unvaccinated-1.5738198

Graham, A. (2021, Mar 17). *Coronavirus: Patios busy, student neighbourhoods quiet on St. Patrick's Day in London, Ont.* Global News. https://globalnews.ca/news/7703616/st-patricks-day-london-ont-2021-coronavirus/

Green, M. J., Croskerry, P., & Rieck, R. (2020). Web Exclusive. Annals Graphic Medicine - Bed Blocker. *Annals of Internal Medicine*, 172(11), W142–W148. https://doi.org/10.7326/G20-0001

Habib M. (2011, Oct 13). *Occupy Canada rallies spread in economic 'awakening'.* Canadian Broadcasting Corporation News. Occupy Canada rallies spread in economic 'awakening' | CBC News

Kannan, S., Shaik Syed Ali, P., & Sheeza, A. (2021). Omicron (B.1.1.529) - variant of concern - molecular profile and epidemiology: a mini review. *European Review for Medical and*

Pharmacological Sciences, 25(24), 8019–8022. https://doi.org/10.26355/eurrev_202112_27653

Khandia R, Singhal S, Alqahtani T, Kamal MA, El-Shall NA, Nainu F, Desingu PA, Dhama K. (2022). Emergence of SARS-CoV-2 Omicron (B.1.1.529) variant, salient features, high global health concerns and strategies to counter it amid ongoing COVID-19 pandemic. *Environ Res*. doi: 10.1016/j.envres.2022.112816. Epub ahead of print. PMID: 35093310; PMCID: PMC8798788

Kolnes, N. H., Eikeland, S. N., Ersdal, T. A., & Braut, G. S. (2022). Estimating the consequences of a COVID-19 super spreader: A stochastic model of a night on the town. *Scandinavian Journal of Public Health*, 50(1), 111–116. https://doi.org/10.1177/14034948211031400

Li, X. P., Ullah, S., Zahir, H., Alshehri, A., Riaz, M. B., & Alwan, B. A. (2022). Modeling the dynamics of coronavirus with super-spreader class: A fractal-fractional approach. *Results in Physics*, 34, 105179. https://doi.org/10.1016/j.rinp.2022.105179

Lord C. (2022, Feb 2). *Trucker convoy: Trudeau says protest 'becoming illegal' as demands for action grow*. Global News. https://globalnews.ca/video/8587818/like-were-being-held-hostage-ottawa-residents-frustrated-as-truck-protesters-refuse-to-leave

McGrail, K. M., Evans, R. G., Barer, M. L., Sheps, S. B., Hertzman, C., & Kazanjian, A. (2001). The quick and the dead: "managing" inpatient care in British Columbia hospitals, 1969-1995/96. *Health Services Research*, 35(6), 1319–1338.

Reston M & Liptak K. (2021, Jan 9). *The day America realized how dangerous Donald Trump is*. CNN Politics. https://www.cnn.com/2021/01/09/politics/donald-trump-dangerous-capitol-riot/index.html

Steinbeck J. (1939). *The Grapes of Wrath*.

Styrborn, K., & Thorslund, M. (1993). Delayed discharge of elderly hospital patients--a study of bed-blockers in a health care district in Sweden. *Scandinavian Journal of Social Medicine*, 21(4), 272–280. https://doi.org/10.1177/140349489302100407

Tasker JP. (2022, Jan 29). *Thousands opposed to COVID-19 rules converge on Parliament Hill*. Canadian Broadcasting Corporation News. https://www.cbc.ca/news/politics/truck-convoy-protest-some-key-players-1.6332312

Chapter 25

The Arrogance of "But All You Need Is a Good Index Finger"

My PSA[553] story is a happy story, though it should not have been. Indeed, I should not even have a PSA story, as I routinely refused my family physician's annual recommendation for routine PSA screening. My ritual reason for my routine refusal was that PSA screening was not covered under our supposed single-tiered health system, and as I teach social justice to medical students, I must walk the talk. A less social-justice–based reason for my routine refusal may have been the certainty in a senior physician's arrogant assurance at the 2010 Canadian Academy of Health Sciences (CAHS) Annual Conference, "But all you need is a good index finger," in response to the almost-motion that PSA screening should be publicly funded. His certainty in "But all you need is a good index finger" echoed in the enthusiastic applause of the audience of CAHS fellows.

Why should I doubt this almost-unanimous consensus of CAHS fellows? After all, we are elected to this "honorific membership

---

[553] PSA stands for prostate specific antigen.

organization"[554] dedicated to health-policy research "by a rigorous peer-review process that recognizes demonstrated leadership, creativity, distinctive competencies and a commitment to advance academic health science."[555] Why should I doubt this almost-unanimous consensus regarding prostate-cancer screening when my research had focused on preventing women's cancers,[556] and in 2010 focused on the injustices in health promotion inflicted on marginalized women, including not having equal access to routine cancer screening;[557] injustices I wanted discussed at this and future CAHS Conferences? Prostate cancer was not the faintest blip on my radar screen in 2010, and would not be until my close friend Stan's diagnosis five years later.

My PSA story is a happy story only because of Stan's unhappy PSA story, and because of Stan's beckoning me from the spiritverse to finally purchase a PSA test. Stan sat beside me through high school, premeds, and medical school; and remained beside me on long-distance phonelines until my last visit to him just prior to his death. My very high PSA levels were almost identical to Stan's, as were my multiple bone metastases and prostate-biopsy Gleason scores of 9s[558] declaring we had a rare aggressive from of prostate cancer termed "prostatic cancer advanced." Another similarity with Stan was that neither the "good index finger" of Stan's family physician nor the "good index finger" of my family physician, both excellent physicians, were good enough to detect our aggressive prostate cancers, even

---

[554] Frank, Battista, Butler, Buxton, Chappell, *et al.*, 2009, p. 3.

[555] Frank, Battista, Butler, Buxton, Chappell, *et al.*, 2009, p. 3.

[556] Joseph, Rab, Panabaker, & Nisker, 2015; Nisker, 1983, 2007; Nisker, Kirk, & Nunez-Troconis, 1988; Nisker, Ramzy, & Collins, 1978; Nisker & Siiteri, 1981; Ramzy & Nisker, 1979; Vanstone, Chow, Lester, Ainsworth, Nisker, *et al.*, 2012.

[557] Joseph, Rab, Panabaker, & Nisker, 2015; Nisker, 2007, 2012; Nisker, Martin, Bluhm, & Daar, 2006; Vanstone, Chow, Lester, Ainsworth, Nisker, *et al.*, 2012.

[558] Prostate biopsies are rarely assigned 10s.

though Stan and I, as physicians, were likely lavished with additional time by "a good index finger."

There is more to the gift of me still being alive to tell you my happy PSA story than having a PSA test because of Stan's unhappy story; there is the string of exceptionalisms and coincidences due to of my gifted position as a physician-educator. If not for this gifted position I would be dead instead of sharing my happy PSA story. Indeed, I feel compelled to share my happy PSA story perhaps out of survivor guilt,[559] though I will not be a survivor for long, or perhaps as a *mea culpa* to the many Canadian men similarly situated to me in having a too-late diagnosis of aggressive "prostate cancer advanced," but who unlike me are dead because they were not physician-educators. Men like my friend Peter.

Peter's PSA story is a horror story of Gothic dimensions, scripted by the inadequacy in our Canadian health system, in which persons are told to assume that if they do not receive a call from their physician's office declaring an abnormal test result, their test result is normal. Peter had an extremely high PSA level, almost identical to Stan's and mine, but Peter's results were buried deep in his chart in his family physician's office. Because of Canada's "don't call us, we'll call you" family-physician-cost-effectiveness rule, Peter assumed there was no need for concern until he developed bladder urgency and went to see his family physician. Peter's family physician went over his chart and found his damning PSA report from three years previous. Peter was referred to a uro-oncologist who ordered a bone scan that demonstrated metastases the full length of Peter's vertebral column, similar to Stan's and mine. The bone scan was followed by a CAT scan that demonstrated "large volume disease" that occupied much of Peter's pelvis and 30 percent of his liver. Peter died while I was writing this story, but could still be alive if his PSA level had not been buried in his chart in his family physician's office.

---

[559] Akhtar, 2009; Gay, 1988; Hartman, 2014; Niederland, 1981.

Another man similar to Stan, Peter, and me in having the extremely high PSA levels of prostate cancer advanced was Canada's social-justice hero, and would-have-been next prime minister, the Honourable Jack Layton. Mr. Layton was born within a year of Stan, Peter, and me, and, like us his prostate cancer advanced was diagnosed too late. As Mr. Layton's health was declining to its inevitable end, the ever-caring Mr. Layton wanted to reassure the many Canadian men with good-prognosis prostate cancer that he had "something more" than the common form of prostate cancer. Dr. Fred Saad of the National Cancer Institute of Canada emphasized that Layton's aggressive prostate cancer advanced makes it clear that "if it's found late, almost all men will die of their disease."[560] Prostate Cancer Canada has argued for decades that annual PSA screening is imperative for all men over age 50 to prevent the 4,100 deaths from prostate cancer each year in Canada.[561]

Two weeks after my PSA test, bone scans, and prostate biopsies declared my bad prognosis, I assembled my three sons on a sofa in my youngest son's home, a floor above the laughter of my grandchildren playing with just-unwrapped Christmas presents. I gently told my boys what they had sensed from the formalness of this assembly: "I have cancer." I went on to explain that the cancer had started in my prostate gland, but was so advanced that they were not offering surgery, or even chemotherapy or radiation; rather just suppressing things for a while with anti-androgens. I asked my sons to celebrate with me the wonderful life I had been given, largely due to them, and to continue to celebrate with me for the hopefully two more years I was told I might have with them.

My younger sons sat frozen, while my eldest leapt from the sofa, upset with my acceptance of this prognosis. Dave demanded a second opinion for an attempt at curative therapy. I hugged him, soothing,

---

[560] Fidelman, 2011.
[561] Bell, Connor Gorber, Shane, Joffres, Singh, *et al.*, 2014; Prostate Cancer Canada, 2017.

"Death's okay Dave, let's just be grateful for the time we've had together, and the time we still will have together." I assured him I was being well looked after by a uro-oncologist who had been a resident with me, and whose office was just 36 steps from my office. He pushed away my hug with, "I can't accept this Dad, so you must promise to call Dr. Carey, and call him today." Dr. Mark Carey had been one of the first residents with whom I had the privilege of working after I finished my research fellowship. He remains a close friend, well known to my sons. More important to my eldest, Dr. Carey is an oncologist, and had just completed a sabbatical at MD Anderson Cancer Center in Houston, where innovative forms of cancer treatment are always being explored. I promised I would call Mark later in the day.

Before we disassembled, I felt I had to tell my sons that my prostate cancer could have been picked up much earlier if I had had annual PSA testing, and that the only reason I had refused routine PSA testing was that it was not covered by the Ontario Health Insurance Plan (OHIP). I felt I owed my sons this explanation, partly as an apology for leaving them sooner than necessary, but mainly to explain that my having prostate cancer permitted them to have funded PSA screening, and with PSA screening they need not fear that what would happen to me would happen to them.[562] Dave hugged me hard and said, "Dad, I respect your decision not to have something that's not publicly funded; I wouldn't have had it either."

My eldest's request to call Dr. Carey furthered the series of exceptionalisms and coincidences that could only happen for a physician-educator who has had the privilege of mentoring many medical students and residents. Mark was grief stricken as he listened to my lack-of-PSA story, and unfortunately agreed with the uro-oncologist's grim prognosis, and my treatment-restriction to androgen deprivation. Almost as an afterthought, he told me the

---

[562] See Chapter 10, She Lived with the Knowledge.

story of his uncle's "horrendous death" from prostate cancer, and suggested I could have "a better death" if I received radiation to my prostate gland. Mark said he would call a radiation-oncologist friend of his, whose office just happened to be down a long corridor from mine.

The radiation-oncologist emailed me the next day, informing me he had ordered a pre-radiation CAT scan that he would make sure happened before we met the following week. However, I asked that the CAT scan be delayed so I could be with Roxanne in Edmonton, where her father was dying from pancreatic cancer.[563] The radiation-oncologist was concerned about delaying, so I promised I would come back as soon as possible.[564] Immediately on my return, the radiation-oncologist met with me during his lunch break, a privilege of course denied non-physicians, and even denied physicians who do not work in the same teaching hospital. I emphasize this lunch-break consultation in a meek attempt to stress that I did not queue-jump his list of other cancer patients. However, I acknowledge that this is a weak apology for a person who teaches ethics and social justice.

The radiation-oncologist looked with seriousness at my scans on his desk computer, bookended[565] by pictures of his children. He took the time to walk me through my bone scan and CAT scan images, and concurred with my bleak prognosis. He also concurred with Mark's suggestion that I would likely have "a better death" with prostate radiation. The visual display of the vertebral metastases I had known about for weeks somehow made the metastases more personal, and deflated what was left in my helium balloon of hope. In addition to the many obvious vertebral metastases, the radiation-oncologist pointed to a potential metastasis in my left leg and wondered whether the possibility of this being a metastasis might be clarified by a new higher-definition bone scan being researched at Princess Margaret

---

[563] See Chapter 18, Victor.
[564] See Chapter 18, Victor.
[565] See Chapter 10, She Lived with the Knowledge; See Chapter 17, Ruth; Nisker, 2015.

Cancer Centre in Toronto.[566] He felt that "knowing the exact extent of the mets" would be important to "better tailor" the radiation. His phone call to Princess Margaret resulted in an appointment there a few days later.

It was less disheartening being in the waiting room of a "Prostate Cancer Centre" two-and-a-half hours north of my office where no one knew me than being in the waiting room of our Cancer Centre where too many people knew me and said, "Hello" or asked, "Which research project are you collaborating on?" or just lowered their eyes. I had arrived at Princess Margaret early, and while waiting for the "higher-definition" bone scan, I went for a walk along the corridor outside the Prostate Cancer Centre. Almost immediately a former medical student, resident, and mentee ran up to me excitedly and vigorously shook my hand. After our so-nice-to-see-yous, Barry asked, "Which research project are you an investigator in?"[567] I responded, "Well I'm actually here as a volunteer for someone else's research." I saw confusion, then concern, cast a shadow over his face, then bow his head and slump his shoulders. The relationship between a mentee and a mentor can be very powerful.[568] Barry gathered his emotions before encouraging in an urgent voice, "There's a new prostate chemotherapist here from Australia who has innovative ideas, and I'll get you in to see him quickly." I was back at Princess Margaret for a lunch-break consultation within the week.

The Australian chemotherapist had of course already received my records from both my uro-oncologist and radiation-oncologist. The

---

[566] See Chapter 5, Princess Margaret; See Chapter 13, The "Helix of Life" Revisited: DNA in Concrete and Not.

[567] Barry was the physician who handed me the *Lancet* article on BRCA gene mutations (Narod, Feunteun, Lynch, Watson, Conway, *et al.*, 1991) on the airplane on the way to a conference, as described in Chapter 15, The Injustice of Needing Angelina Jolie. Barry is also the physician who helped my Sister in Chapter 10, She Lived with the Knowledge, as explored in greater detail in *Sarah's Daughters* (Nisker, 2012).

[568] Levy, Katz, Wolf, Sillman, Handin, *et al.*, 2004; Nisker, 2003, 2006.

chemotherapist studied my metastases on his laptop and quickly recommended I receive docetaxel chemotherapy. He based his recommendation on an unpublished study in which he was one of the researchers. He then took the time to pull up the PowerPoint slides of his research and show me several graphs demonstrating extension of life expectancy with docetaxel, even with aggressive metastatic prostate cancer advanced. The extent of life-extension was not yet known, but the Australian chemotherapist was confident that half of the patients receiving docetaxel would live an additional two years. However, he could not prescribe me docetaxel, as his research study was completed and docetaxel was not yet offered clinically for prostate cancer.

When I returned home, it occurred to me that if the radiation-oncologist felt a higher-definition research bone scan might "better tailor" his radiation, why not use the power of MRI to even more precisely delineate my pelvic metastases, and better tailor my imminent radiation? I called my radiation-oncologist and asked if he would book an MRI for me. He compassionately soothed, "Jeff, we don't do MRIs for prostate cancer because they're expensive, and bone scans and CAT scans are pretty good." Then he of course placed my name on the MRI cancellation list. I had the MRI within a week, and was back in his office for another lunch-break consultation a few days later.

A close friend and colleague of mine, again a former medical student and resident, sat beside me in the radiation-oncologist's office. Akira had learned of my cancer diagnosis from seeing my name on the patient list the last time I was in our Cancer Centre. He immediately entered the clinic room where I was waiting and hugged me hard. I needed this hug, as I was depleted from having been in our Cancer Centre's waiting room, where patients had waited for me and I found myself surrounded by persons seeming so ill. Akira insisted he be with me during all my future medical visits, emphasizing the juxtaposition of his clinic just below my radiation-oncologist's office, and just above the radiation machines.

As my radiation-oncologist's computer screen lit up, Akira riveted himself on the MRI images of my pelvis. I felt his heart drop and saw his shoulders sag as he recognized before I did a previously unseen thumb-size metastasis replacing my right pubic bone. This large metastasis had been invisible not only to my bone scans but also to the pelvic CAT scan on which my radiation field would have been based. My prognosis was even worse than anticipated. Akira's chin met his chest. He fought back tears. I squeezed his arm.

The radiation-oncologist suggested, "It would be easier to tailor the radiation if the bulky pubic bone metastases was reduced in size a bit with chemotherapy first."

While Akira and I were in the radiation-oncologist's office, Roxanne was at the memorial service for her father, Victor.[569] I regret not being there with her; not being there for her. It was this regret, rather than the verdict of my sooner-than-expected death, that emanated through me in the radiation-oncologist's office. Not that I am courageous; rather I had previously made peace with my death sentence of prostate cancer advanced. I knew Roxanne would want me to convey my grimmer prognosis on the phone to her that day; however, as that day was already a difficult one for Roxanne, I debated whether a day's delay would be okay. As Roxanne would not think a day's delay okay, I called her later that day.[570]

The following week, I met with a chemotherapist in our Cancer Centre. He was aggressively against prescribing docetaxel. He had printed out the PowerPoint graphs kindly sent to him from the chemotherapist at Princess Margaret, and had them spread on his claustrophobic[571] consultation room's desk. He told me that the docetaxel research on prostate cancer advanced "had not yet gone through peer review." The chemotherapist said he was skeptical about prolongation of life with docetaxel for prostate cancer advanced, but

---

[569] See Chapter 18, Victor.
[570] See Chapter 18, Victor.
[571] See Chapter 3, I'm Sorry Ronnie.

was quite sure of docetaxel's major side effects because of his experience with docetaxel for other cancers.

I responded, "I have nothing to lose, and if docetaxel prolongs my life just long enough to see one more grandchild's smile the side effects are worth it."

He was not happy with this response, and went into a long lecture about his obligation as a physician to his Hippocratic Oath: "*Primum non nocere*, first do no harm." He then stressed the many harms of docetaxel, stood up, and said it was his right to refuse to give docetaxel to me. Of course I argued with him, which he of course was not used to. My closing argument was, "I teach informed choice to medical students and residents, so if there's anyone who can understand and appreciate the risks of docetaxel with unproven benefit, it's me." I quickly added, "I would be happy to sign an Against Medical Advice form." He was perturbed, but stared further at the PowerPoint graphs, then stared at the computer-images of my MRI's thumb-size metastasis replacing my right pubic bone. He took a deep breath. "Okay, but just because you're a physician here."

I believe my large pubic bone metastasis, which was only identified by my "expensive" MRI, in addition to me being a physician in the same teaching Hospital, encouraged the chemotherapist to call the Hospital's Chief Pharmacist and arrange to have docetaxel made available for me. She was reluctant, but after meeting with me eventually agreed. A year later the research proving docetaxel's life-extension in prostate cancer advanced was published in the *New England Journal of Medicine*.[572]

Let us return to "But all you need is a good index finger." The "good index finger" of more than one good physician did not detect my prostate cancer advanced, nor Stan's prostate cancer advanced, nor Peter's prostate cancer advanced, nor Jack Layton's prostate cancer advanced. In addition, my family physician always took considerable extra time examining my prostate gland because she

---

[572] Sweeney, Chen, Carducci, Liu, Jarrard, *et al.*, 2015.

knew I was going to refuse her recommendation of purchasing a PSA test, and also because I just happened to have been her "Faculty Mentor" when she was a medical student decades earlier. Even with her additional time commitment, her "good index finger" could not detect any abnormality in my prostate gland that had spawned many metastases, including the thumb-size metastasis in my right pubic bone.

My uro-oncologist's even more experienced "good index finger" could not detect with certainty my prostatic cancer, even though he was aware of my very high PSA level. He told me my prostate gland was firm but not irregular or enlarged, and reserved the diagnosis of cancer until the results of the ultrasound-guided prostate biopsies that he had already ordered for me came in. Similarly, the "good index finger" of the radiologist who performed the "random biopsies" of my prostate gland could not detect prostate cancer. I stress random, as not only was no cancer apparent with his "good index finger," no cancer was apparent with his real-time ultrasonography, just a "dense" prostate gland. These lacks of "good index finger" detections were due to the fact that even though my prostate cancer was very aggressive, my prostate gland was neither enlarged nor nodular; rather homogeneously filled with the aggressive prostatic cancer, cancer that was demonstrated on all 10 random biopsy specimens.

So why at the 2010 Canadian Academy of Health Sciences Annual Conference were the leaders of the Canadian medical establishment so full of the arrogance of "But all you need is a good index finger"? This arrogance included non-physician members of CAHS, such as biostatisticians, epidemiologists, and health-policy pundits, who argued that PSA is not a cost-effective test, as "most prostate cancers are slowly progressive and not life threatening."[573] However, "most prostate cancers" do not include my form of prostate cancer, nor Stan's, nor Peter's, nor Jack Layton's, nor that

---

[573] Bell, Connor Gorber, Shane, Joffres, Singh, *et al.*, 2014.

of the 4,100 Canadian men who die each year from prostate cancer.[574] Although statisticians are willing to accept these deaths in the name of cost-effectiveness, this acceptance is bad medicine and sets a dangerous precedent in Canada's supposedly single-tiered health system; a precedent that Canada's wealthy men routinely bypass by purchasing PSA tests.

A similar example of statisticians accepting deaths in the name of cost-effectiveness is their recommendation that women should stop examining their breasts, because women finding benign lumps leads to unnecessary ultrasounds and needle biopsies.[575] However, their cost-effective statistics ignore the individual woman who finds a lump in her breast and, after confirmation by her family physician, is sent for the ultrasound-guided biopsies that reveal breast cancer. That individual woman's breast self-examination may have saved her life, or at least given her more happy years. Although statisticians have contributed greatly to the health promotion of Canadians, they do not have the privilege of touching the trust inherent in the handshake of a person in their care. Thus, their proclamations can easily be distracted by population-based cost-effectiveness statistics, without appreciation of the impact of their proclamations on individual persons.

As family physicians appreciate that their index fingers are not inviolate, not even "good," they routinely recommend the annual purchase of a PSA test to men over 50. Though statisticians deem this test not cost-effective, most family physicians continue to favour their responsibility to the individual person sitting in their office over cost-effectiveness statistics. This responsibility is being threatened by emphasis on family physicians becoming "gatekeeper/managers."[576] In addition, few front-line family physicians are ever elected Fellows

---

[574] Bell, Connor Gorber, Shane, Joffres, Singh, et al., 2014; Prostate Cancer Canada, 2017.
[575] Klarenbach, Sims-Jones, Lewin, Singh, Theriault, et al., 2018; Tonelli, Connor Gorber, Joffres, Dickinson, Singh, et al., 2011.
[576] Dath, Chan, & Abbott, 2015.

of the Canadian Academy of Health Sciences, as most family physicians are so immersed in caring for individual persons that they rarely have time for the research and national health-policy work required to qualify them.

My PSA story is a happy story, and always will be a happy story, even when my life ends in a few years. I write my happy PSA story to lobby for public funding of PSA screening, so that all men in Canada can have a happy PSA story through early detection of their prostate cancers. Even men with "prostatic cancer advanced" can have a happy PSA story if their PSA test reveals their aggressive prostate cancer early enough. Prostate Cancer Canada has actively advocated for public funding of PSA tests for many years and continues to advocate provincial governments to cover PSA tests.[577] The arrogance of senior physicians in believing "But all you need is a good index finger," combined with the callousness of cost-effectiveness statistics, still inhibits public funding of PSA screening in some provinces. Further, lack of public funding diminishes the importance of PSA screening; after all, if PSA screening was important, it would of course be publicly funded. In addition, socio-economically disadvantaged men cannot afford to purchase access to early detection of their prostate cancers.

The initiation of public funding of annual PSA tests for all men over 50 in Canada would make my happy PSA story an even happier one.

Portions of a shorter preliminary version of this chapter were published in *Medical Ethics* in 2020.[578]

---

[577] Prostate Cancer Canada, 2017.
[578] Nisker, 2020.

References

Akhtar, S. J. (2009). *Comprehensive dictionary of psychoanalysis.* Karnac.

Bell, N., Connor Gorber, S., Shane, A., Joffres, M., Singh, H., Dickinson, J., Shaw, E., Dunfield, L., Tonelli, M., & Canadian Task Force on Preventive Health Care. (2014). Recommendations on screening for prostate cancer with the prostate-specific antigen test. *Canadian Medical Association Journal, 186*(16), 1225–1234.

Dath, D., Chan, M.-K., & Abbott, C. (2015). *CanMEDS 2015: From manager to leader.* The Royal College of Physicians and Surgeons of Canada. www.royalcollege.ca/rcsite/documents/cbd/canmeds-2015-manager-to-leader-e.pdf

Edwards S. (2021, Dec 27). Coronavirus Update: COVID-19 cases in Canada top two million as hospitals brace for Omicron wave's unknowns. Globe and Mail. https://www.theglobeandmail.com/canada/article-coronavirus-update-covid-19-cases-in-canada-top-two-million-as/

Fidelman, C. (2011, August 24). Jack Layton's letter gives cancer patients hope. *Montreal Gazette.* https://montrealgazette.com/news/jack-laytons-letter-gives-cancer-patients-hope

Frank, C., Battista, R., Butler, L., Buxton, M., Chappell, N., Davies, S. C., Edwards, A., Henshall, C., Joly, Y., Jordan, G., Kealey, T., Wolfson, M. C., & Woolf, S. H. (2009). *Making an impact: A preferred framework and indicators to measure returns on investment in health research.* Canadian Academy of Health Sciences. https://cahs-acss.ca/making-an-impact-a-preferred-framework-and-indicators-to-measure-returns-on-investment-in-health-research/

Gay, P. (1988). *Freud: A life for our time.* Norton.

Hartman, J. J. (2014). Anna Freud and the Holocaust: Mourning and survival guilt. *International Journal of Psychoanalysis, 95*(6), 1183-1210.

Joseph, M., Rab, F., Panabaker, K., & Nisker, J. (2015). Feelings of women with strong family histories who subsequent to their breast cancer diagnosis tested BRCA positive. *International Journal of Gynecological Cancer 25*(4), 584-592.

Klarenbach, S., Sims-Jones, N., Lewin, G., Singh, H., Theriault, G., Tonelli, M., Doull, M., Courage, S., Garcia, A. J., Thombs, B. D., & Canadian Task Force on Preventive Health, C. (2018). Recommendations on screening for breast cancer in women aged 40-74 years who are not at increased risk for breast cancer. *Canadian Medical Association Journal, 190*(49), E1441-E1451.

Levy, B. D., Katz, J. T., Wolf, M. A., Sillman, J. S., Handin, R. I., & Dzau, V. J. (2004). An initiative in mentoring to promote residents' and faculty members' careers. *Academic Medicine, 79*(9), 845-850.
https://www.ncbi.nlm.nih.gov/pubmed/15326007

Narod, S. A., Feunteun, J., Lynch, H. T., Watson, P., Conway, T., Lynch, J., & Lenoir, G. M. (1991). Familial breast-ovarian cancer locus on chromosome 17q12-q23. *Lancet, 338*(8759), 82-83.

Niederland, W. G. (1981). The survivor syndrome: Further observations and dimensions. *Journal of the American Psychoanalytic Association, 29*(2), 413-425.

Nisker, J. (2006). A covenant model for the medical educator-student relationship: Lessons from the covenant model of the physician-patient relationship. *Medical Education, 40*(6), 502-503.

Nisker, J. (2012). *From Calcedonies to Orchids: Plays promoting humanity in health policy.* Iguana Books.

Nisker, J. (2015). *Patiently waiting for...* Iguana Books.

Nisker, J. (2020). Arrogance of "but all you need is a good index finger": A narrative ethics exploration of lack of universal funding of PSA screening in Canada. *Journal of Medical Ethics, 46*(4), 249-252.

Nisker, J., Martin, D. K., Bluhm, R., & Daar, A. S. (2006). Theatre as a public engagement tool for health-policy development. *Health Policy, 78*(2–3), 258–271.

Nisker, J. A. (1983). Screening for endometrial cancer. *Canadian Family Physician 29,* 961–965.
https://www.ncbi.nlm.nih.gov/pubmed/21283374

Nisker, J. A. (2003). Medical students mirror and hold mirrors. *Journal of Obstetrics and Gynaecology Canada, 25*(12), 995–999.

Nisker, J. A. (2007). The need for public education: "Surveillance and risk reduction strategies" for women at risk for carrying BRCA gene mutations. *Journal of Obstetrics and Gynaecology Canada, 29*(6), 510–511.

Nisker, J. A., Kirk, M. E., & Nunez-Troconis, J. T. (1988). Reduced incidence of rabbit endometrial neoplasia with levonorgestrel implant. *American Journal of Obstetrics & Gynecology, 158*(2), 300–303.

Nisker, J. A., Ramzy, I., & Collins, J. A. (1978). Adenocarcinoma of the endometrium and abnormal ovarian function in young women. *American Journal of Obstetrics & Gynecology, 130*(5), 546–550.

Nisker, J. A., & Siiteri, P. K. (1981). Estrogens and breast cancer. *Clinical Obstetrics and Gynecology, 24*(1), 301–322.

Prostate Cancer Canada. (2017). Statistics.
http://www.prostatecancer.ca/Prostate-Cancer/About-Prostate-Cancer/Statistics

Ramzy, I., & Nisker, J. A. (1979). Histologic study of ovaries from young women with endometrial adenocarcinoma. *American Journal of Clinical Pathology, 71*(3), 253–256.

Sweeney, C. J., Chen, Y. H., Carducci, M., Liu, G., Jarrard, D. F., Eisenberger, M., Wong, Y. N., Hahn, N., Kohli, M., Cooney, M. M., Dreicer, R., Vogelzang, N. J., Picus, J., Shevrin, D., Hussain, M., Garcia, J. A., & DiPaola, R. S. (2015). Chemohormonal therapy in metastatic hormone-sensitive prostate cancer. *New England Journal of Medicine, 373*(8), 737–746.

Tonelli, M., Connor Gorber, S., Joffres, M., Dickinson, J., Singh, H., Lewin, G., Birtwhistle, R., Fitzpatrick-Lewis, D., Hodgson, N., Ciliska, D., Gauld, M., Liu, Y. Y., & Canadian Task Force on Preventive Health, C. (2011). Recommendations on screening for breast cancer in average-risk women aged 40–74 years. *Canadian Medical Association Journal, 183*(17), 1991–2001.

Vanstone, M., Chow, W., Lester, L., Ainsworth, P., Nisker, J., & Brackstone, M. (2012). Recognizing BRCA gene mutation risk subsequent to breast cancer diagnosis in southwestern Ontario. *Canadian Family Physician, 58*(5), e258–e266. https://www.ncbi.nlm.nih.gov/pubmed/22734169

Chapter 26

Our Third COVID Summer

In our third COVID summer
"COVID fatigue" sets in
So politicians lift restrictions
Before Public Health gives permission
As it appears in politicians' best interests
To soothsayer[579] COVID is "in the rear-view mirror"
A falsehood to foster "Open for Business"[580]
And summer-pleasure for future voters

---

[579] A "soothsayer" in Ancient Greek Mythology predicted the future often based on the throwing of small bones like dice or the observing of smoke plumes from their throwing water on hot stones

[580] The word "Business" was used frequently by Business-elected Ontario Premier Doug Ford in slogans like "We're open for business". I capitalize "Business" because Ford presents it as a deity we need to worship.

In our third COVID summer
The calm before autumn's COVID prediction
Has become fodder for ignoring
Public Health's dire warning
As we bask in "catch-up" celebrations
That lack masks and social distance
Applauded by our relaxed premier
Because our ICUs can handle more COVID

In our third COVID summer
Our premier applauds
Our "high vaccination rate"
As an excuse to remove the mask mandate
But COVID vaccines are not potent
And ICUs no place to live in
Or die in as many will in the autumn
With or without vaccination

In our third COVID summer
Variants breed beneath the surface
Not causing symptoms severe enough
For hospital admissions
Thus evidence for statisticians
To convince Business-elected politicians
To insist on masks and social distance
To diminish autumn's COVID

In our third COVID summer
I fear my grandchildren now consider me
The masked Poppa Jeff
Standoffish for some reason
As I no longer lift them in the air
To kiss their squealing belly-buttons
Nor let them trampoline my abdomen
To spring their gorgeous giggles

In our third COVID summer
Rather I see confusion in my grandchildren's eyes
When they see my eyes above a mask
Instead of seeing the me who loves them dearly
But whose immunity is diminished
By chemotherapy's gift
Of extra years with grandchildren's laughter
Though COVID diminishes another summer

In our third COVID summer
The removal of mask mandates
From most spaces in our nation
Causes COVID-complacency
But also COVID-paranoia for the older
Including me when I face breathing on my masked face
Or the mask-straps on my neck
At maskless clusters of check-out counters

In our third COVID summer
My triple-vaxxed Brother called
To let me know he has COVID
And was wasting his summer with cough
And on CBC[581] television I learned
Our triple-vaxxed Prime Minister has COVID
As does the triple-vaxxed prime minister of the United Kingdom
And the triple-vaxxed president of the United States

In our third COVID summer
A proliferation of COVID-aggression[582]
Spawns placards in bank windows
ABUSE OF TELLERS WILL NOT BE TOLERATED
But there are no signs on roads that warn
DANGER: COVID-AGGRESSIVE DRIVERS
Who cut in and out of traffic
And are more dangerous than the COVID virus

In our third COVID summer
I also observe COVID-aggression when I run
Close to the curb as per usual
But now hear profanity hurled
Through car windows speeding by
Or amplified at stop signs
Both with the addition of a certain finger
Though we're in COVID together

---

[581] Canadian Broadcasting Corporation.
[582] COVID was claimed as the reason for more aggressive behaviours of males, including a dramatic increase in the incidence of rape (Canadian Broadcasting Corporation National News Aug 2, 2022).

In our third COVID summer
"Dare I eat a peach"[583] of patio dining
With servers no longer masking
My answer is "Yes please"
For we need to go on living
While hoping "opening" won't increase dying
Before winter moves servers indoors
And clusters more maskless COVID

In our third COVID summer
"Big Pharma" promises new vaccines
Guaranteed to finally work
But we've heard this optimism before
And Omicron's variants will likely morph faster
Than scientists can pursue them
And will definitely spread faster
When there are no masks for inhibition

As colder weather suggests autumn
I let my guard down like many others
Before winter insists further reclusiveness
Which will again force me from our Hospital
And its intimacy of "in-person"
Meetings and teachings and learnings
Onto Zoom which is less human
By Chi-square significance

---

[583] TS Elliot, 1915.

References

Elliot TS. (1915). *The Love Song of J. Alfred Prufrock*. Poetry: A Magazine of Verse at the instigation of Ezra Pound.

Appendix I

The Psych Experiment

I was fortunate to be one of the 100 high school students bestowed direct entry into the six-year medical school program at the University of Toronto, the last year direct entry was permitted. With the decision to dump direct entry, the powers that be dumped the 100 of us for two "elective" courses into the general student population, one of which was Psych 100, the most heavily enrolled course in the most heavily enrolled university in Canada.

Over 1500 students were crammed into the mammoth Psych 100 course, which had a mammoth crammable textbook, *Psychology*, to cram, with over 900 double-columned, small-print pages. The lectures for Psych 100 took place in a mammoth amphitheatre that I actually entered twice; the restriction of my attendance was based on the grading system that restricted the information spouted from the lectern to only 20 percent of the overall grade. The rest of one's grade came from tests on the mammoth text, and on a smaller and more interesting text, *Psychobiology*, which being grounded in the scientific method was more convincing to students raised in math and physics.

The prof of Psych 100 was a youngish energetic guy who had just arrived at the University of Toronto from an Ivy League university, and was already famous because he had invented an innovative evaluation system he termed "The Chunk System." In the Chunk System, evaluation focused intensively on a small portion of the mammoth material on which you graded yourself prior to the tests, with the understanding that your eventual grade would be extrapolated from your performance on the pre-graded section. His evaluation strategy was based on the reasonable presumption that a student would be consistent in their knowledge of all sections. I did not take umbrage with the Chunk System method of examination, nor with its youngish prof; however, I did take umbrage with his insistence that all students must "volunteer"[584] for one "psych experiment."

At 0900 the Saturday morning of the psych experiment, I dutifully reported as directed to an old house just south of New College, the college where they housed and fed med students. The house was locked. I knocked repeatedly to no avail. Obviously I was there at the wrong time, or on the wrong day, and either way I would not be returning. Just as I turned to leave, a diminutive, very beautiful woman at whom I had gazed longingly daily over a table in New College Library the prior month came up the steps. She softly asked if the door was locked. I opened my mouth to respond but could not, as I was too busy silently thanking the spiritverse for my good fortune. Finally, I exhaled, "Yes."

Much too soon a guy in his late twenties with a goatee, probably a grad student, opened the door and took us to separate rooms. I was handed a sheet of paper with instructions as to what was required of me during the experiment. I was to act like a typical jock, which was not hard for me, and use "superior strength to resist push of partner." I again thanked the spiritverse for the gift I was about to receive, as "push of partner" meant this lovely woman would be touching me. After what seemed like an eternity, the goateed grad student opened a second door to my confining room and led me into an obviously

---

[584] To "volunteer" one must have the right to refuse participation without retribution (Nelson et al, 2008; Nisker, White et al, 2006; Vanstone et al, 2012; Nisker, 2013: Morgenstern et al., 2015; Wada & Nisker, 2015).

one-way-mirror room in which the brief encounter[585] of our psych experiment would occur. My lovely partner, whom I would be asking out on a date for that evening, was already there.

My partner's eyes looked at the floor as I wondered if she could hear my pounding heart's accelerating beating over my rapid breathing. We heard the grad student's sanctimonious voice on an intercom say, "Start." My partner pushed the entirety of her small body repeatedly into my chest, soon getting out of her sweet breath. Too soon the intercom said, "Thank you." I also said "Thank you," hoping her eyes would lift to mine, but they did not. Instead the goatee came rushing in and led us back to the rooms in which we were previously sequestered. He handed me a heavy-lead pencil, along with a multiple-choice questionnaire, asked me to fill it out, fold it, and "push it into the cardboard box as if it was a ballot box." As one had to be 21 to vote in those days, the analogy with a ballot box was lost on me. He added I could leave as soon as it was completed. As the questionnaire was quite lengthy, I quickly ticked the boxes without reading the questions and ran out the front door to wait for my Saturday night date.

I waited. I waited. I waited.

I assumed that my partner's questionnaire was longer than mine, or more likely that she was taking the questions more seriously than I had. After a half hour of waiting, I began thinking the house might have another door, and I ran around to the back in panic. I found a door there, but not the alluring student. I ran back to the front door only to find a classmate of mine waiting to enter to "volunteer" for his obligatory psych experiment. There did not seem to be a partner there for him. We waited together for a while before I dejectedly went

---

[585] *Brief Encounter* is a film directed by David Lean in 1945, with the screenplay written by Noël Coward, based on his 1936 one-act play *Still Life*. The central characters are a gentle young woman and a pleasant man, who just happens to be a physician, who have tea together in a train station every Thursday across a small table. The couple gaze longingly at each other before they must board different commuter trains home to their families.

home, encouraged only by the knowledge that I would see my lovely partner on Monday in New College Library.

I thanked the spiritverse all weekend in anticipation of Monday. I planned to stay in New College Library all day until my date for next Saturday night, and maybe for the rest of my life, came in. The classmate I saw waiting to go in for his psych experiment came in, and much to my chagrin sat where my date always sat. I told him I was waiting for the love of my life to sit where he had just sat, and asked him to move several seats away. A few hours later, when we took a sandwich-machine break, I whispered to him my good fortune with the psych experiment partner and began describing her in almost intimate detail. Suddenly he stopped me with, "Wait a minute, you're describing that real bitch who was *my* psych experiment partner."

I felt a flush of injustice flash through my core. I had been betrayed by my professor, not to mention by my partner, who I finally realized was a co-conspirator. My Pollyanna ethos assured me that she was coerced by the prof, likely her grad-school supervisor, to serve in her conspiratorial role. The prof had used her physical appearance, not to mention physical actions, to make the psych experiment work.

The feelings of injustice pumping from my heart made it easy to find fault with my prof, not only founded in the deception of the psych experiment, but in the power differential that gave me no choice but to "volunteer" for his research; research that had no benefit for me,[586] except to pass his course, but that had significant

---

[586] Many medical research studies, particularly the randomized clinical trials approved by Research Ethics Boards, may have clinical benefits for participants, as well as for patients in the future. However, there is less benefit in early phases of medical research, so the participants must undergo a very lengthy "informed consent" process, and most supply renumeration for their participation. This system of encouragement to participation by financial renumeration is highly problematic as it will more likely recruit socio-economically disadvantaged "volunteers" (Nisker et al, 2006).

benefit for him. The research papers emanating from the psych experiment would earn the prof praise, and possibly even promotion with tenure. In addition I have no recollection of signing an informed consent document for research participation, but I may have. If I did sign such a document, I did so without reading it carefully, but I would have remembered the word "deception" if it had been mentioned.

Of course the word "deception" could not be mentioned in the informed consent document, or the psych experiment would not have induced me to be enamoured with my partner, and thus it would not have worked the way the prof wanted. As the combination of "deception" and "informed" is an oxymoron, there was motivation for exclusion of the word "deception" if indeed an "informed" consent document existed. In checking with classmates, none had recollection of the word "deception." Similarly, none of my classmates remembered a clause stating that "you have volunteered for this experiment without coercion." Bringing deception to their attention consistently evoked feelings of being reduced to cannon fodder by a professor who had power over us.

Of course none of us blamed our lovely partner, even though we all agreed she must be in on the deception. I went so far as to contend that the psych experiment must be part of her work as the prof's graduate student, or perhaps research assistant, and thus her graduation or salary was dependent on complying with the prof's expectations. I argued that she was just as disempowered as we were, likely even more. No doubt my late-teen androgens played a part in convincing me of this argument. I should have asked the few women in my medical school class if they had had a similar experience with the same partner, or if they had an attractive male partner, or if they were even included as participants in the psych experiment.

At New College Library's once-quiet central table, other med students who had "volunteered" to participate in the psych experiment were expressing anger at being duped. I assured them I would find a way to get back at our prof. It did not take long to come up with the strategy. As our now infamous prof was famous for developing the Chunk System method of evaluation, it was obvious I should target the Chunk System. Later that day, after careful examination of how the Chunk System functioned, I discovered a fatal flaw that could be exploited to condemn the sanctity of our prof's innovation.

For the Chunk System to work as a rigorous method of examination of over a thousand double-columned small-print pages, it assumed that each student would read at least most of the pages, even if only superficially. This assumption was not unreasonable, as many of the more than 1500 students in Psych 100 probably considered psychology their future career. However, neither the Chunk System nor the prof had considered that a group of multiple-choice–expert medical students, compelled to take Psych 100, would work as a team to divide up the many pages, crib them down to likely multiple-choice questions, and share their crib notes with classmates.

About 30 of us came together to crib the massive material; some in order to achieve an A, others like me to get back at our deceptive prof. Our cribbing made the psych exam even easier than anticipated. A week later, we observed our destruction of the Chunk System in the bell curve[587] of the exam results, which, rather than being a bell curve, had a distorting blip of 30 marks between 90 and 95 percent. The vast majority of the more than 1500 students in the course were devastated, because hardly any of them received As, and, again, many of them wanted to be psychologists. I actually went to the lecture theatre I had so carefully avoided to see if there would be any debris discussed in the debriefing. There was, along with tears, and questions about not understanding how a grade could be so low when a student had studied so hard.

---

[587] A bell curve is a stats method.

I reflected on the psych experiment several times over the years, particularly while writing papers on informed choice.[588] It consistently occurred to me that psychology research must have a lower bar for informed consent than medical research to achieve Research Ethics Board approval. This perception was validated when serving on Western University's Senate, where I observed strong pushback from the Department of Psychology regarding the recent merger of our University's Social Science and Medical Science Research Ethics Boards. Comments from non-medical scientists were along the lines of, "Why should the social sciences be held to the same standards as medical science research?" Other comments included, "Before the two research ethics boards were merged there was never a problem."

I had taken umbrage with the existence in a well-respected university in a well-respected liberal democracy of a system that compelled students to "volunteer" to participate in a psych experiment in order to pass a course. I contend our recruitment for participation was coercive, rather than the "voluntary" participation essential for informed choice in research, no matter the stripe of the research. Perhaps the psych experiment was part of the reason I eventually became an ethics researcher, particularly one interested in informed choice.[589]

I reflect briefly on the psych experiment in the "The Cement Spiral"[590] as I run past the house in which it occurred, and also in Chapter 13 of this book, "The 'Helix of Life' Revisited: DNA in Concrete and Not."

---

[588] Nisker, White et al, 2006; Vanstone et al, 2012; Nisker, 2013: Morgenstern et al., 2015; Wada & Nisker, 2015.
[589] I use the word "choice" rather than "consent," as "consent" seems to mean to me that "choice" has already occurred or ought to occur, rather the remaining just an option (Nisker, White et al, 2006; Vanstone et al, 2012; Nisker, 2013: Morgenstern et al., 2015; Wada & Nisker, 2015).
[590] Nisker, 2018.

References

Morgenstern J, Hegele RA, Nisker J. (2015). Simple genetics language as source of miscommunication between genetics researchers and potential research participants in informed consent documents. Public Underst Sci, 24 (6): 751-66.

Nisker, J., White, A., Tekpetey, F., & Feyles, V. (2006). Development and investigation of a free and informed choice process for embryo donation to stem cell research in Canada. Journal of obstetrics and gynaecology Canada, 28(10), 903–908. https://doi.org/10.1016/S1701-2163(16)32279-4

Nisker J. (2013). Informed choice and PGD to prevent "intersex conditions". The American journal of bioethics : AJOB, 13(10), 47–49. https://doi.org/10.1080/15265161.2013.828125

Nisker J. (2018). The Cement Spiral. Journal of obstetrics and gynaecology Canada, 40(6), 643–645. https://doi.org/10.1016/j.jogc.2018.01.003

Vanstone, M., Kinsella, E. A., & Nisker, J. (2012). Information-sharing to promote informed choice in prenatal screening in the spirit of the SOGC clinical practice guideline: a proposal for an alternative model. Journal of obstetrics and gynaecology Canada, 34(3), 269–275. https://doi.org/10.1016/S1701-2163(16)35188-X

Wada, K., & Nisker, J. (2015). Implications of the concept of minimal risk in research on informed choice in clinical practice. Journal of medical ethics, 41(10), 804–808. https://doi.org/10.1136/medethics-2014-102231

Acknowledgements

Mariko Obokata

Jennifer E. L. Ryder

Katharine Timmins

Roxanne Mykitiuk

About the Author

Jeff Nisker is a physician, researcher, and writer. Through his plays, short stories, and poems, readers and theatregoers are immersed in the inequities of new scientific capacities. His plays have been performed throughout Canada, and in the United States, the United Kingdom, Australia, and South Africa. Six of his plays were published in *From Calcedonies to Orchids: Plays Promoting Humanity in Health Policy*. Jeff was profiled in the *Canadian Medical Association Journal* in an article titled "Theatre of Social Justice."

    Jeff's numerous research and education awards include Canada's Royal Conservatory of Music's Excellence in Education Award, which "recognizes the efforts of an outstanding educator who embraces the idea that the arts have the capacity to change the world." Jeff has also received many other awards for his innovative research and education initiatives, including Western University's Scholar Award for Innovation in Research and Education, and the Society of Obstetricians and Gynaecologists of Canada's President's Award for the most significant contribution to the specialty. In 2020, Jeff continued to win awards for innovation in research and education. Jeff has also co-held a Canadian Institutes of Health Research/Health Canada grant to research public engagement and citizen deliberation for health-policy development through his innovative use of full-length theatre. Jeff has served on the editorial boards of *Journal of*

Medical Humanities and *Ars Medica,* and was the international representative on the Board of the Centre for Literature and Medicine. Jeff was chosen by the Canadian Broadcasting Corporation's Peter Gzowski as one of the 13 "Best Minds of Our Time."

Dr. Jeff Nisker
LHSC – 800 Commissioners Road East, Room E2-620E
London, ON N6A 5W9
jeff.nisker@lhsc.on.ca
Office: 519 685-8781
Home: 519 430-6475

www.ingramcontent.com/pod-product-compliance
Lightning Source LLC
Chambersburg PA
CBHW031422150426
43191CB00006B/362